Dados Internacionais de Catalogação na Publicação (CIP)
(Câmara Brasileira do Livro, SP, Brasil)

Kleinubing, Maximilian
The flight test organizations : improving efficiency and safety during aircraft development / Maximilian Kleinubing ; [tradução Andrew Charles Fox]. -- 2. ed. -- Porto Alegre, RS : Ed. do Autor, 2025.

Título original: As organizações de ensaios em voo.
Bibliografia.
ISBN 978-65-01-62380-1

1. Aeronaves 2. Aeronáutica 3. Engenharia I. Título.

25-291458 CDD-629.13

Índices para catálogo sistemático:

1. Aeronáutica : Tecnologia 629.13

Eliete Marques da Silva - Bibliotecária - CRB-8/9380

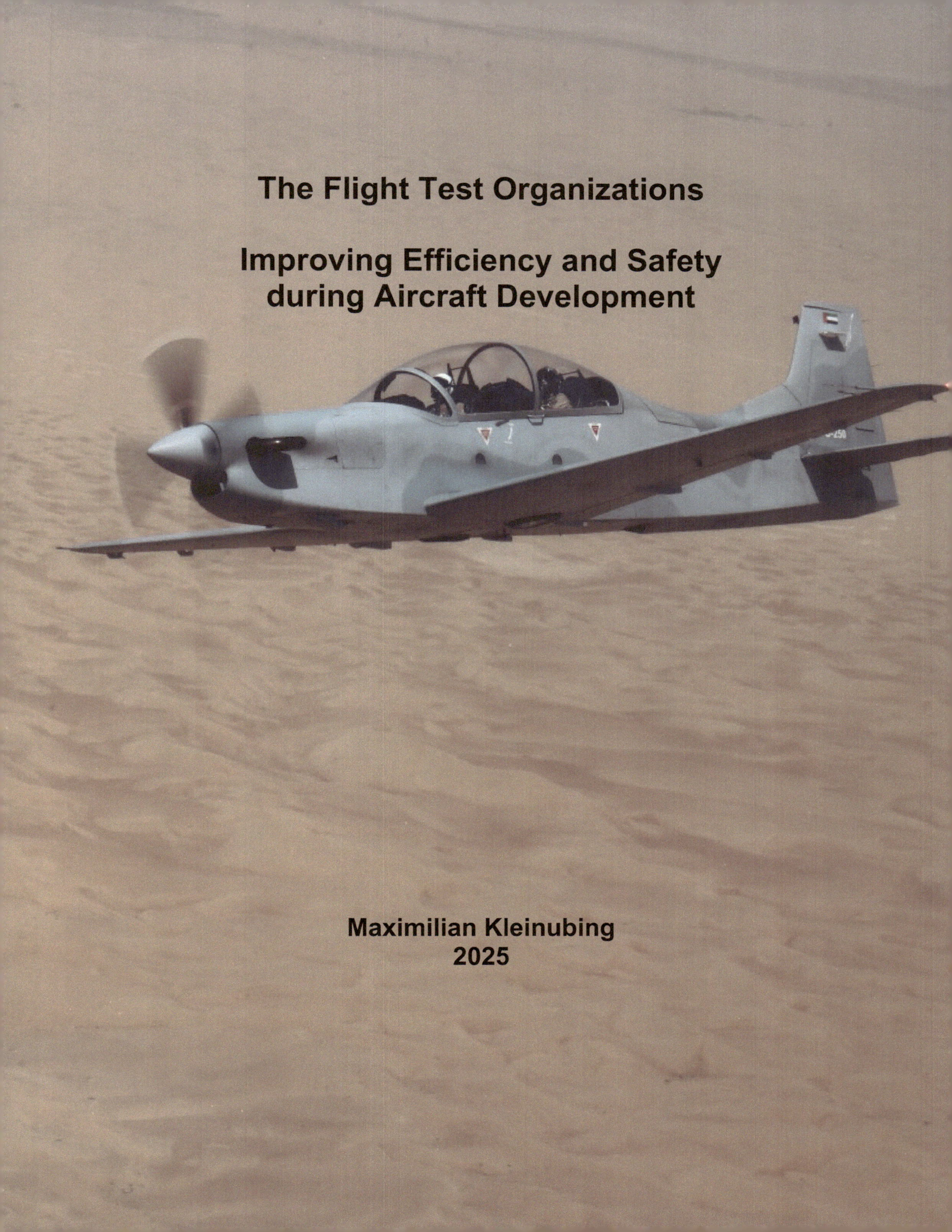

The Flight Test Organizations

Improving Efficiency and Safety during Aircraft Development

Maximilian Kleinubing

2025

To my parents Vera and Sadi, who gave me, with so much love, the gift of dreams.

THANKS

This project could not have been completed without the direct or indirect support of the many people I mention here.

My wife, Léa, for helping me to balance family and my work on this book.

My family, for always encouraging my achievements and adventures.

My work colleagues at Calidus Aerospace, for contributing so much by discussing issues related to this book. Especially my friends Lahej Al Falasi, Nilson Perini, Ramiro Chagas de Carvalho, Celso Carvalho, Jose Rodolfo Rodrigues, Weber Barbosa, Qaisar Majeed and Romulo Sotrate.

My fellow engineers and pilots, who have been with me for 25 years. In particular, my friends Paulo Quattrocchi, Roberto Zanasi Jr, Fabrício da Silva Pereira, Márcio Brisola Jordão, Glen Peach, Marcelo Basile, Clodoaldo Matias, Gilberto Meira Cardoso, John Sevalho Corção, Alexandre Matta, Aírton Manoel Rodrigues, Richard Lucht, Artêmio Relvas de Almeida, Fernando Benitez Leal, Thaís Frank Cruz, Alexandre Mellek, Luiz Ricardo Noce, Gustavo Moraes Cazelli, Márcio Panssiera, Daniel Pisani Dias, Clênio Sobreira, Leonardo Gutierrez, Adriano Paulo Villas Boas, Robert McCausland, Nicholas Coop, Martin Ascher, Wendy Ng, Matthew Hartkop, Patrick Hubert, Daniel Rodrigues Ribeiro, and André Luiz Forni.

The tutors at the Technological Institute of Aeronautics (ITA), the School of Advertising and Marketing (ESPM), and the University of São Paulo in São Carlos (USP São Carlos) for motivating me to research and to keep studying. In particular, Fernando Catalano and Paulo Celso Greco Jr.

The Navy History and Heritage Command, the Smithsonian National Air and Space Museum and the Lockheed Martin Company for their kind support with historical images used in this book.

Calidus Aerospace, represented by its CEO, Dr. Khalifa Al Blooshi, for their invaluable support during the difficult time when illness and uncertainty were part of my life.

Flight Test Organizations

Improving Efficiency and Safety During Aircraft Development

Editora Marathon
2nd Edition

SUMMARY

CHAPTER 1. FLIGHT TEST ORGANIZATIONS 12
CHAPTER 2. THE FLIGHT TEST DOCTRINE 26
CHAPTER 3. FLIGHT TEST PROCESSES 74
CHAPTER 4. FLIGHT TEST OPERATIONS MANUAL 102
CHAPTER 5. FLIGHT TEST ENGINEERING 110
CHAPTER 6. FLIGHT TEST PILOTS 118
CHAPTER 7. FLIGHT OPERATIONS SUPPORT 126
CHAPTER 8. FLIGHT TEST INSTRUMENTATION 130
CHAPTER 9. MAINTENANCE 140
CHAPTER 10. SAFETY MANAGEMENT SYSTEM (SMS) 154
CHAPTER 11. ORGANIZATIONAL FACTORS AND RISKS 166
CHAPTER 12. RISK MANAGEMENT IN FLIGHT TESTING 172
CHAPTER 13. CASE STUDIES 188
CHAPTER 14. AERONAUTICAL PRODUCT DEVELOPMENT 222
CHAPTER 15. AERONAUTICAL DESIGN ORGANIZATIONS 228
CHAPTER 16. AERONAUTICAL PRODUCTION ORGANIZATIONS 236
CHAPTER 17. FINAL WORDS 244
CHAPTER 18. REFERENCES 248

FLIGHT TEST ORGANIZATIONS

I

"Every soldier must master the bow before the sword."
Genghis Khan

BEFORE YOU START

This book provides ideas on how a Flight Test Organization should function to best achieve its objectives, and:

- Avoid accidents
- Avoid mistakes
- Develop the team
- Have the best tools

Despite their similar purposes, there are major differences between Flight Test Organizations, such as culture the number of people involved the number of aircraft under their responsibility the number of annual flight hours the academic and technical training of their teams, and the country they operate in.

While acknowledging these differences, this book seeks to identify and detail the common activities among these organizations. They have not only activities in common, but structure (see Chapter 1) operating doctrines (see Chapter 2) processes (see Chapter 3) the need for training (see Chapter 3) the need to have a Risk Management System (see Chapter 12) and to comply with the Organization's Risk Management System procedures (see Chapter 10) They are part of an Aeronautical Design Organization (see Chapters 15), have a Maintenance team (see Chapter 9), an Instrumentation team (see Chapter 8), and a Flight Test Operations Manual (see Chapter 4). Finally, they can also work with an Aircraft Production Organization (see Chapter 16).

I believe that understanding these characteristics allows a Flight Test professional to better perform their duties in the organization and, therefore, be more effective.

The information and techniques in this book are based on professional experience and grounded in repetition and analysis. They are not simply ideas, but methods based on real experiences.

INTRODUCTION

Flight testing is like an industry that operates within the aviation industry worldwide. This industry is made up of hundreds of organizations that have a similar knowledge base and culture. Supporting this knowledge are training schools that train new professionals in this field and provide courses that help professionals specialize in certain types of tests, or aircraft, and keep up to date.

This industry generates thousands of jobs worldwide and has been expanding since the year 2000 due to growing demand for new types of aircraft (eVTOL, UAVs, new propulsion systems, and so on). Every new aircraft design must be developed and tested. Most also require some kind of certification by government bodies (FAA, EASA, ANAC, and so on) that represent the public interest in having safe means of transportation (without, however, worrying about the commercial validity of these products). Certification is an important step that manufacturers pursue to make their products more attractive on the market and, consequently, more saleable.

Flight Tests are mostly carried out by professional organizations specializing in this activity, known as Flight Test Organizations (FTOs).

A lack of available data means the number and type of Flight Test Organizations in the world is impossible to estimate.

TYPES OF FLIGHT TEST ORGANIZATION

There are numerous types of Flight Test Organization (FTO) around the world. In general, Flight Test Organizations carry out test flights on prototype aircraft, or aircraft modified for testing. To better understand how each organization operates, it is important to divide them according to their main characteristics.

Flight Test Organizations can be divided according to:
- Their purpose
- Whether they are civilian, or military
- Type of aircraft being developed.

Types of Flight Test Organization

PURPOSE

a) Research and development

These are Flight Test Organizations belonging to Research and Development Bodies, funded by companies and governments. Many Flight Test Organizations belong to government research bodies. Examples include:

1. NASA Research Institutes (Edwards, Langley, etc.)
2. Deutsches Zentrum für Luft- und Raumfahrt (DLR)
3. Netherlands Aerospace Center (NLR)
4. Edwards Air Force Base
5. Flight Test Research Institute (of the Brazilian Ministry of Aeronautics)
6. Eglin Air Force Base (Specializing in Weapons Systems Research and Testing)
7. NATO RTO (NATO Research and Technology Organization)

These organizations set out to be at the forefront of new technologies, making their work important for the future of the industry as a whole. Their results often become benchmarks and are eventually used by aeronautical companies around the world.
The cutting-edge nature of Research and Development Organizations makes their operation very delicate, and different practices need to be used. Historic projects such as the X-1, the Apollo Program, the Space Shuttle Program, and dozens of others, were and still are developed in organizations of this type. These organizations' work teams,

modes of operation, equipment, techniques, and tools must be of a superior standard, highly efficient, and technological.

Some sources define this activity as Experimental Flight Testing.

b) Teaching

These are flight test schools whose purpose is to train new engineers and test pilots. To this end, they usually have both fixed-wing and rotary-wing aircraft. These aircraft usually have instrumentation that enables detailed analysis of the data, as well as serving as tools for aerial instruction. They are:

1) IPEV Flight Test School, São José dos Campos, Brazil
2) US Air Force Flight Test School (USAFTPS), Edwards, CA, USA
3) US Navy Flight Test School (USNTPS), Patuxent River, BA, USA
4) National Test Pilot School (NTPS), Edwards, CA, USA
5) International Test Pilot School (ITPS), London, ON, Canada
6) Sirrus Testing School (Sirrus TPS), São José dos Campos, Brazil
7) Royal Test Pilot School (RAFTPS), Boscombe Down, United Kingdom
8) École du Personnel Navigant d'Essais et de Réception (EPNER), Istres, France
9) Fedotov Test Pilot School, Zhukovsky, Russia
10) Soviet and Russian Air Force Test Pilots Training Center, Akhtubinsk, Russia
11) Indian Air Force Test Pilot School, Bangalore, India
12) Test Flying Academy of South Africa, Oudtshoorn, South Africa
13) Neo Energy Aviation Academy, Berlin, Germany
14) Euro Flight Test, Koblenz, Germany
15) Escuela Española de Ensayos en Vuelo y Aeronavegabilidad, Madrid, Spain.

It is in these organizations that the foundations of Flight Test culture worldwide are formed. They establish the basic concepts and tools that will be used in other organizations. Their objectives are also aimed not only at imparting technical knowledge but also at shaping the attitudes and discipline of future test crew members.
The different courses offered by these schools are usually divided into four types:

- Type I: this is a complete Flight Test course that presents and demonstrates all the concepts in the three main areas Flight Testing is divided into: Performance, Flight Qualities, and Systems. This type of course usually includes an initial section on mathematics and physics, which provides the necessary reinforcement of the students' knowledge. The class load is always heavy, combined with an overload of test flights, each followed by detailed Flight Test reports. This type of course usually lasts a year, with a workload of around 800 to 1,000 lesson hours, added to around 100 hours of rehearsal flights. This borders on the impossible, and many participants drop out or are terminated because they cannot keep up with the course. In addition, cost is severely restrictive, at around 600,000 to 800,000 dollars for pilots and 300,000 to 500,000 dollars for engineers.

- **Type II**: this is a short Flight Test course that presents and partially demonstrates the concepts of the three major areas of Flight Testing: Performance, Flight Qualities, and Systems. It usually focuses on civil aircraft systems and certification. Most last around six to eight weeks. Students also carry out test flights to demonstrate techniques and concepts. This type of course is significantly less costly and easier for students to follow. Because of its good cost-benefit ratio, it is widely used in the industry.

- **Type III**: this is a condensed introductory flight test course that introduces general concepts in the areas of flight testing. It usually lasts two weeks and is either purely theoretical or has a few demonstration flights. This type of course is introductory and quite simple to teach compared with the other types.

- **Type IV (Short Courses)**: These are usually one-week courses that aim to provide a quick overview of a specific topic so that it can be applied quickly by the students. They are usually for members of the aeronautical industry who are looking for better qualifications in each area.

The descriptions above show the great difference between course types.

However, despite all the effort required at Flight Test schools, it is a mistake to think that a professional coming from these organizations is ready to take on major responsibilities in other types of organizations, where the characteristics, processes, responsibilities, and objectives are radically different. Flight test schools provide the right *background* from which, with constant effort and dedication, a flight test professional will develop. They can accelerate and foster this growth but do not replace the actual practice of the activity.

c) Aircraft manufacturers

These are organizations belonging to private companies that develop and market their aeronautical products. Virtually all aircraft manufacturers have their own Flight Test Organization, as this is essential for the development of new aeronautical products.
It is here, in a more random and often chaotic testing environment, that the practices developed in Research Organizations and disseminated in Teaching Organizations are repeated exhaustively, facing many practical problems in product development. The challenge here is to apply known concepts in a professionally (and culturally) diverse cultural environment. Often, what works well in teaching and research organizations fails inexplicably. On the other hand, procedures that are considered incomplete or flawed can be adopted here to simplify and reduce costs, or for reasons related to a lack of resources, or training. Factors such as cost, time, and the availability of skilled labor and resources become decisive.

CIVILIAN OR MILITARY ORGANIZATIONS

Civil

Civilian Flight Test Organizations (FTOs) are usually part of for-profit companies that seek to achieve profit through their products and services. FTOs must, therefore, work in harmony with a larger organization and get results and make deliveries for the rest of the organization as if they were a service provider. They usually have an associated Design Organization (DO) and Production Organization (PO). See Chapter 15 for more details on Aeronautical Design Organizations (DOs) and Chapter 16 on Aeronautical Production Organizations (POs).

Military

Military Flight Test organizations are not-for-profit, and their objectives are defined by the armed forces they belong to. These organizations follow standards or practices defined by the Armed Forces and usually do not follow civilian standards (or they partially follow them).

It is common for military Flight Test Organizations not to have an associated Design Organization and Production Organization. They work with a limited production of test bodies made for Flight Test purposes.

The absence of a Design Organization and a Production Organization, together with other characteristics inherent to the armed forces (rigid hierarchy, military employment, highly disciplined personnel with a linear, rising career path), makes military organizations operate in a significantly different way from civilian organizations.

Moving from one type of organization to another can, for these reasons, be quite difficult for the professional and sometimes takes a significant amount of time (several years).

TYPES OF AIRCRAFT

Fixed-wing aircraft

They make up the majority of Flight Test Organization aircraft in the world, given the existence of a large associated market.

Rotary-wing aircraft

They account for a smaller part of Flight Test Organization aircraft in the world, given the smaller associated market in terms of quantity, but which is significant in terms of investments.

VTOL and e-VTOL aircraft

VTOL aircraft are those capable of taking off and landing vertically. e-VTOL aircraft are electrically propelled aircraft capable of taking off and landing vertically.

They make up an expanding part of Flight Test Organization fleets in the world given the growing associated market. Most of these are start-ups that are still trying to consolidate their products. The market for this type of aircraft is still being consolidated. It is estimated that there are around 288 Flight Test Organizations for VTOL and e-VTOL aircraft worldwide. [1]

Unmanned Aircraft (Drones)

These are an expanding part of Flight Test Organizations in the world given the growing associated market. It is estimated that there are around 1,076[2] unmanned aircraft flight test organizations around the world. This market has been developing fast due to growing commercial and military interest. Recent armed conflicts, such as the Ukrainian War, which began in February 2022 when Russia invaded Ukraine, have shown the effectiveness of military use of unmanned aircraft.
On the other hand, the use of anti-drone weapons has opened up a new operational need within the mission of military aircraft.

FLIGHT TEST ORGANIZATION STRUCTURES

All Flight Test Organizations need to achieve their objectives:

- Qualified people
- Aircraft
- Infrastructure (hangars, flight equipment, maintenance equipment, etc.)
- Processes
- Results
- Deliveries

An example of a very efficient organizational chart for a Flight Test Organization is shown below.

[1] According to www.trackxn.com
[2] According to the website www.droneii.com in its Drone Market Map

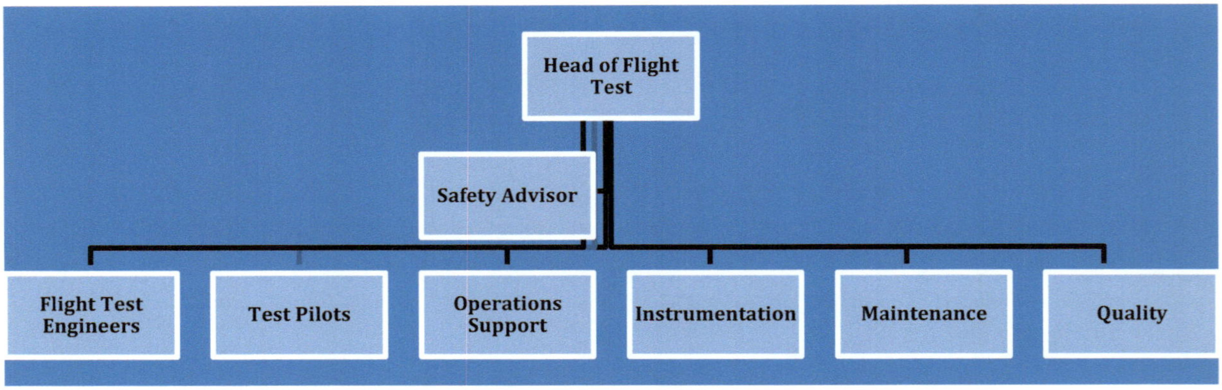

Typical organization chart of a Flight Test Organization

PEOPLE

A Flight Test Organization team must be made up of the following professionals:

- Test Pilots
- Flight Test Engineers
- Instrumentation engineers and technicians
- Maintenance engineers and technicians
- Quality engineers and technicians

In addition to these professionals, others are usually needed to support processes, coordinate activities, and facilitate the organization's necessary deliveries. They are:

- Flight controllers
- Designers (for FTI systems, mechanical and electrical systems, structures, etc.)
- Operating Engineers
- Engineers and IT technicians
- Administrative assistants
- Corporate security
- Cleaning and conservation assistants

AIRCRAFT

Of course, aircraft are a fundamental part of Flight Test Organizations, and much of their attention and effort is focused on them.
The aircraft is divided into the following types

- Fixed wing
- Rotating wing
- VTOL
- e-VTOL
- Drones

Aircraft are also divided according to their application:

- Proofs of Concept: aircraft designed to prove the viability of a project. They usually have characteristics that are not yet known, such as new engines, new materials, or new aerodynamic concepts.
- Prototypes: these are aircraft that are partially representative of a production aircraft. Sometimes their propulsion and fuel systems are representative but usually, there are one or more systems that are not representative of a production aircraft.
- Production aircraft: aircraft produced with all the design features incorporated for delivery to customers.

INFRASTRUCTURE

Infrastructure is one of the specialized and necessary items for FTOs. It has a marked impact on the results achieved and can be limiting, or expansive. We can mention:

OPERATIONS INFRASTRUCTURE

- Airports
- Hangars (air-conditioned, or not)
- Offices
- Briefing rooms
- Auditoriums
- Maintenance rooms
- Yards for engine turning
- Yards for compass calibration
- Operations control and support towers
- Parking
- Small hangars (for flight clearance)
- Flight Equipment Rooms
- Flight equipment

INFRASTRUCTURE FOR INSTRUMENTATION

- Sensor calibration laboratories
- Telemetry rooms
- Mobile telemetry
- Telemetry data systems
- FTI stock (cameras, sensors, cables, tools, data acquisition systems)

INFRASTRUCTURE FOR INFORMATION TECHNOLOGY

- IT rooms (servers)
- Computer and software maintenance laboratories
- IT stock (computers, displays, cables, data storage systems)

MAINTENANCE INFRASTRUCTURE

- Hydraulic sources
- Electrical sources
- Pneumatic fountains
- Anemometer sources
- Cranes
- Tire shop
- Toolmaking
- Prototyping room
- Equipment and cabling assembly room
- Maintenance stock (parts, tools, measuring instruments)

PROCESSES

The Flight Test processes will be dealt with in detail in Chapter 3.

RESULTS

A Flight Test Organization's results differ by type of organization. However, some results are common to all of them:

- Preparing aircraft for flight
- Modification of aircraft to test new configurations and improvements
- Test Flights
- Instrumented Aircraft
- Installed and calibrated instrumentation parameters
- Data recorded during test flights
- Videos and images recorded during test flights
- Flight Test Plans
- Test flight reports
- Flight Test Data Reduction Tools

DELIVERIES

It is important to differentiate between deliverables and results. Results can be achieved without a formal handover. This happens all too often, and it is not ideal. Formalizing results guarantees recognition of the work and the spread of good practices throughout the organization. For deliverables to be satisfactory, their content and form must be appropriate for their intended audience. It is common for people to want to format their deliverables to achieve particular objectives not aligned with the broader objectives of the organization, generating distortions.

Let me give you an example: suppose the organization has a weekly report on its aerial activities, tests, and results obtained. However, the head of a particular sector, for example, the head of Maintenance, wants to include a summary of their activities in this report for the company's top management. This is an example where adding new information can change the intention of the report and make it unnecessarily long, compromising its effectiveness.

THE FLIGHT TEST DOCTRINE

2

"Direction is more important than speed."
Richard L. Evans

At the heart of all Flight Test Organizations is a very similar working philosophy. This is commonly referred to as doctrine. The term doctrine means a set of good practices designed to increase efficiency and safety.

The doctrine aims to answer questions such as:

- What can you do to avoid accidents?
- How to respond to failures and emergency situations?
- What kind of resources and safety equipment are needed?
- What to prioritize to protect the crew and aircraft?

It is the result of experience acquired and matured over a long period and must always be taken into account.

The doctrine is of little relevance regarding:

- Social conventions
- Pride
- Power, careers, and so on

When there is urgency (motivated by danger) in making a decision, we do precisely those things that are instinctively ingrained in our minds.
Other things are forgotten, or modified, in the high emotional charge of the moment. Another characteristic of this type of situation is that they occur in a very short period, sometimes a few seconds, which limits the time for analysis and decision-making, sometimes to fractions of a second.

It is important to show the importance of the doctrine in activities that involve decision-making in high-stress situations, as can occur during Flight Trials.

PRE-EXISTING CONDITIONS

The following conditions can always be considered true in any project:

- Project: There are always flaws
- Human error: always present
- Flaws: Always there
- Processes: Always flawed
- Deadlines: Always very short
- Working hours: Often on the edge
- Software: Always contains bugs
- Better to act than react
- Example: worth more than 1,000 words
- A picture is worth 1,000 words.

Flight tests are carried out by teams made up of people from different backgrounds but with a common goal.
Teams have opponents with similar objectives in companies spread across various parts of the globe.

Flight Test Team

On the other hand, teams must deal with the difficult task of human coexistence, and the quality of this coexistence significantly affects the outcome of teamwork and, therefore, the survival of the team.

Coexistence is facilitated by politeness, patience, tolerance, and humility and hindered by arrogance and prejudice. A great team must be very strong technically, but even stronger in terms of cooperation.

Flight testing is NOT usually characterized by speed and repeatability (although it does pursue them). It is characterized by a great deal of preparation work, rapid responses, understanding of results, and failure analysis.

DOCTRINE IN THEORY

When it comes to doctrine, theory is just as important as practice. So, here are the main concepts.

Flying is an unnatural activity.
Possible remedies for this feature are:

- Repeat until the unnatural becomes natural
- Anticipate the next step
- Know where to stop.

Best attitudes to save time and avoid arguments:

- Do not jump to conclusions
- Seek first to understand and then to be understood
- Do not complicate things if you can simplify them
- Predict, test, and compare results

DOCTRINE IN PRACTICE

- Do not create anything that has not been planned and described. If in doubt, leave it for another day.
- Discuss any doubts before the test.
- Make checks against information received.
- Study often. This is irreplaceable.
- Know the test subject very well.
- Study the limitations of the aircraft in detail.
- Discuss safety analysis with others.
- Note the main conclusions of the safety analysis.
- Discuss it with the pilot or engineer who will be carrying out the flight.
- Use a document with Aircraft Operational Limitations (AOL). This document guarantees the safety of the Test Campaign. Ensure that it is up to date and contains clear, applicable information.
- Prepare a Test Engineer's Manual. This document is a compendium of information about an aircraft or program, organized in a clear and easily accessible way for briefings and use during tests. It should always be up to date and contain clear, useful information.
- Put the main limits applicable to the test in the Test Order. Do not trust that all AOLs will be remembered during the test.
- Try to find out the Lessons Learned about that type of test in previous campaigns in any other program.
- Read about the type of test, testing techniques, applicability, etc.
- Use Resource Review (RR), Technical Review (TR), and Safety Review (SR) meetings to mature the test plan. Arrange these meetings as soon as possible.
- Check that the sequence of points is the best from the point of view of weight/CG, configuration, procedures, test area, and aircraft energy.

- Know what safety equipment will be used and its operation.
- Include mechanics and instrumentation staff in the briefings and debriefings.
- Prepare mentally for the flight and try to find flaws in the planning.
- Prepare for the test briefing.
- Make the briefing concise and operational. The briefing is not a lecture and should not use audiovisual resources that distract attention from the procedures to be carried out or make it too long. The ideal length for a briefing is 45 minutes.
- Study the applicable requirements and the Test Proposal. This enables a better assessment of the validity of the results.
- Estimate the speeds that will be used during the test. This avoids possible errors during the test.
- Plan the time spent and fuel used during the test.
- Carry out concise debriefings with your impressions of the flight and the results (expected compared with achieved). The ideal length of a debriefing is up to 45 minutes.
- Concern yourself with the certification and future operation of the aircraft, relating your observations to its mission.
- Try to identify possible impacts of the flight results on future aircraft tests and record them in the Test Report so that the whole team can be aware of them.
- Make a concise Test Report with the pilot/engineer who carried out the test.
- Make sure that the recorded data has been validated by the applicant after the test.
- Qualitative data and impressions are very important. Recorded data will not be able to provide information that only those who were in the aircraft during the test have. Do not assume that "everything is being recorded."
- Make sure corrective actions are being taken for problems encountered.

EXAMPLES OF DOCTRINE

To exemplify the best Flight Test doctrine practices, this section brings together some of the most significant work in the history of aviation (see below).

THE TEAMWORK OF CHUCK YEAGER AND JACK RIDLEY

Here is an account from Brigadier Charles "Chuck" Yeager.

"By far the smartest guy in the class was a small, very thin bomber pilot called Jack Ridley, from Oklahoma. He had studied at Caltech and was ten steps ahead of the others and much more than that ahead of me. The course was very difficult, and it took me a while to get to grips with it, because I lacked the theoretical part. We were taught to reduce data, represent graphs and maps, using calculus and algebra formulas. When I didn't understand something, I asked Ridley for help, and he always explained the problem in a way that was clear to me. As it turned out later, Jack would do this for me again and again, saving my life on many occasions."[3]

Dick Frost was the person who taught Yeager about the X-1 aircraft. But regarding the characteristics of the airplane, he trusted the advice of Jack Ridley. Ridley was the best student of the famous Hungarian aerodynamicist Theodore von Braun during his studies at Caltech. Colonel Boyd told him he could trust Ridley's opinion and that's what he did.

During the X-1 flight operations Yeager would sit behind the B-29 pilots in an apple crate. When the time to launch the X-1 approached he would go down the stairs followed by the co-pilot, Jack Ridley. He would slide down into the X-1 cockpit and close the hatch. At this moment, without power, the cockpit of the X-1 would fall into complete darkness.

Later in the program the engineers found out that a shock wave was been formed on the X-1 elevator hinge line at high mach numbers. This would make the pilot to lose the longitudinal control of the aircraft. The whole team thought that was the end of the program. But Ridley sat on a chair at the end of the table writing some calculations and finally said "maybe he can fly using the horizontal stabilizer". Bell engineers had put an actuator on the horizontal stabilizer because they foresaw the possibility of shock waves there. That approach made it possible to achieve and finally fly at supersonic speeds.

Ridley made important contributions to the team that broke the sound barrier for the first time in history of aviation. He was a pilot and engineer and the trust and partnership between him and Yeager was decisive to the success of the program.

---

[3] *Yeager, an Autobiography*, General Chuck Yeager and Leo Janos.

JAMES DOOLITTLE'S LEADERSHIP

Born in 1896 in the city of Alameda in California, the boy showed himself to be an engineer from an early age, assembling and disassembling household appliances to the amazement of his family. Watching an aircraft take off at an aeronautical exhibition in Seattle when he was 13, he knew he wanted to fly.

He studied Mechanical Engineering at the University of California at Berkeley. From then on, he also began training as an aviator at a nearby airfield.

When the First World War came along, he enlisted in the US Army Signal Battalion and was soon sent for training. When he was sent to France to operate the DH-4 bomber, he was squadron leader and commander. His daring and innovative tactics made him a point of reference among his fellow airmen. In addition, he was credited with shooting down several enemy aircraft, putting him on the list of the best American aces of the First World War.

When he returned to the United States after the war, he continued his studies and graduated from the famous Massachusetts Institute of Technology (MIT) as an aeronautical engineer in 1925.

Air racing became a well-known sport in the 1920s, and the Schneider Trophy became the most famous air race in the world. The trophy was created in 1912 by French industrialist and aviator, Jacques Schneider, to encourage the development of seaplanes. The aviator who completed the course in the shortest time won the competition. Doolittle won the 1925 Schneider Trophy flying a Curtis R3C-2 and achieving an average speed of 201 knots.

On September 4, 1922, he made the first coast-to-coast flight in the USA, flying a De Havilland DH-4 from Jacksonville, Florida, to San Diego, California, with a total flight time of 21 hours and 19 minutes. The Army awarded him the "Distinguished Flying Cross" medal for this achievement.

Doolittle was the first pilot to fly "blind," that is, using only the aircraft's on-board instruments and with no view of the outside world. One of his most notable achievements also included testing flight instruments, notably the artificial horizon. Many years later, at an advanced age, when asked in an interview which of his achievements he considered to be the most important for aviation, he cited the development of instrument flight.

Doolittle taxiing the R3C-2 in Bay Shore Park.[4]

After the surprise attack to Pearl Harbor, President Franklin D. Roosevelt wanted to give a response as early as possible and bring the war to Japan.

The so-called "Doolittle Raid." was a surprise bomber attack on Japan in response to the Japanese attack on Pearl Harbor and was the first attack carried out inside Japanese territory in the Second World War.

The idea of attacking Japan using medium-range bombers came from Navy Captain Francis S. Low, who observed twin-engined aircraft training at an airfield with the design of an aircraft carrier deck painted on the runway.

With great support from General Henry “Hap” Arnold, Chief of Staff of Army Air Forces, James Doolittle, famous at the time for his achievements in Army aviation was called in to plan the attack.

The mission's objectives were to fly some 2,400 nautical miles and carry 2,000 pounds of weaponry to carry out a bombing raid over Japan. At the time, there was no aircraft capable of carrying out such a mission from a land-based airfield, so an aircraft carrier would have to be used.

The B-25 Mitchell was chosen, a medium-range bomber, to carry out the mission. The B-25 was capable of carrying this 2,000 lb load (the equivalent of four 500 lb bombs each) but had a range of no more than 1,300 nautical miles. To achieve the mission objective, the B-25 would have to be modified to reduce weight and drag and, in

[4] https://www.usni.org/magazines/naval-history-magazine/2021/february/curtiss-racers

addition, it would have to carry more fuel. The most important modifications were the installation of a 160-gallon neoprene folding auxiliary fuel tank attached to the top of the bomb bay and additional fuel cells to increase the fuel capacity from 646 to 1,141 US gallons (2,445 to 4,319 liters).

Also, replacement of the Norden bomb sight, the most advanced aiming equipment of the time, with an improvised sight developed by pilot Capt. Ross Greening, which was nicknamed the "Mark Twain" sight. The reason for this substitution was to prevent the Norden sight, one of the secrets of the war, from falling into enemy hands.

The pilots were recruited from the 17th Bomber Squadron, the first to receive the B-25s and one of the most experienced in the bombing mission.

The 24 aircraft were sent to Eglin Air Base for three weeks of training. They were made up of volunteers recruited by Doolittle, but none were Navy pilots. A Navy pilot, Lt. Henry Miller, was sent there to train them in carrier take-off and other naval aviation tasks.

Maintenance problems and accidents, however, ended up affecting the amount of training. However, three planes had to be discarded. The first was due to an accident during landing, the second because it suffered serious damage in an accident during take-off, and the third was discarded due to a shimmy in the nose that could not be repaired in time.

In the end, there would be five crew members per aircraft, for a total of 16 aircraft and 80 crew members.

On April 1, 1942, the 16 B-25s were loaded onto the USS Hornet along with their crews and maintenance teams.

On April 2, 1942, the USS Hornet set sail on its historic mission to Japan. A few days later, the Hornet would rendezvous with the aircraft carrier USS Enterprise and its escort of cruisers and destroyers. The total force comprised seventeen ships, including two aircraft carriers.

Besides the task force, two submarines were sent to patrol stations to report information about Japanese naval forces activities.

It was only during the two-week journey across the Pacific to Japan that Doolittle assembled his team and gave them all the details of the mission.

The plan was to launch the B-25s when they were 450 NM from Japan, to allow them to reach the Chuchaw base in China after the bombing.

B-25 Mitchells aboard the USS Hornet bound for Japan[5]

On the morning of April 18, 1942, about 650 NM off the Japanese coast, the convoy was spotted by a Japanese fishing boat, which radioed that the fleet was approaching. The boat was sunk by the cruiser Nashville and part of its crew was captured.

Doolittle and the commander of the USS Hornet, Captain Marc Mitscher, gave the command to launch the B-25s, even though they were 170 NM further away than planned. The first to take off was Doolittle, at a take-off distance of 142 m. Reports of the take-off say that the airplane came very close to hitting the water but managed to start climbing in time. Then all the others took off, too.

Flying in groups of three planes, they proceeded to their objectives flying just above the waves to avoid radar detection. It took about four hours to fly to Tokyo. In all, ten targets were attacked in Tokyo - two in Yokohama, and the others in Yokosuka, Nagoya, Kobe, and Osaka.

They encountered a few enemy planes and some anti-aircraft artillery, but none were shot down. After the attack, the planes headed for the east coast of China, except for aircraft 16, which flew towards Russia due to lack of fuel. It turned out that the aircraft's

[5] U.S. Navy Photo Courtesy of Naval History and Heritage Command

carburetors had been adjusted for maximum efficiency in low-altitude flight, but by mistake, the two carburetors on this particular aircraft were replaced with stock carburetors in Sacramento, without Doolittle's knowledge. The aircraft ended up making a forced landing some 64 km north of Vladivostok.

The crew members later reported that, despite the bad weather on the way to Japan, they were lucky enough to find a good tailwind after the attack, which made it possible for almost everyone to reach China as planned. After about 13 hours of flight, all the B-25s had made forced landings, or the crew had parachuted out of the planes. Only one of the crew died when he abandoned the aircraft.

Doolittle and his crew parachuted over China and were helped by Chinese soldiers and peasants. Most of the crew who hit China were also rescued by Chinese soldiers or peasants, and eventually made it to safety.

However, the Chinese ended up paying dearly for helping the Americans. Many were arrested and executed by the Imperial Japanese Army. In all, it is estimated that around 250,000 Chinese were killed during the search for the Americans.

All the B-25s, except for aircraft 16, which landed in Russia, were lost. Of the 80 crew members, 64 escaped alive, eight were captured by the Japanese, of whom three were executed, and one died in prison. Two of the missing crew drowned when their aircraft made a forced landing in the sea. Their remains were brought to the USA after the war and buried with military honors in the Golden Gate National Cemetery.

The eight soldiers who were arrested by the Japanese were put on a starvation regime. They were sentenced to death, but five of them had their sentences reduced and three were executed. Those who remained in prison were released by American troops in August 1945. Four Japanese officers were tried and convicted of war crimes. They were sentenced to hard labor for three to five years, with one of them sentenced to nine years.

Doolittle and his team's example of courage and leadership served as a beacon for all who followed in the struggle during the war.

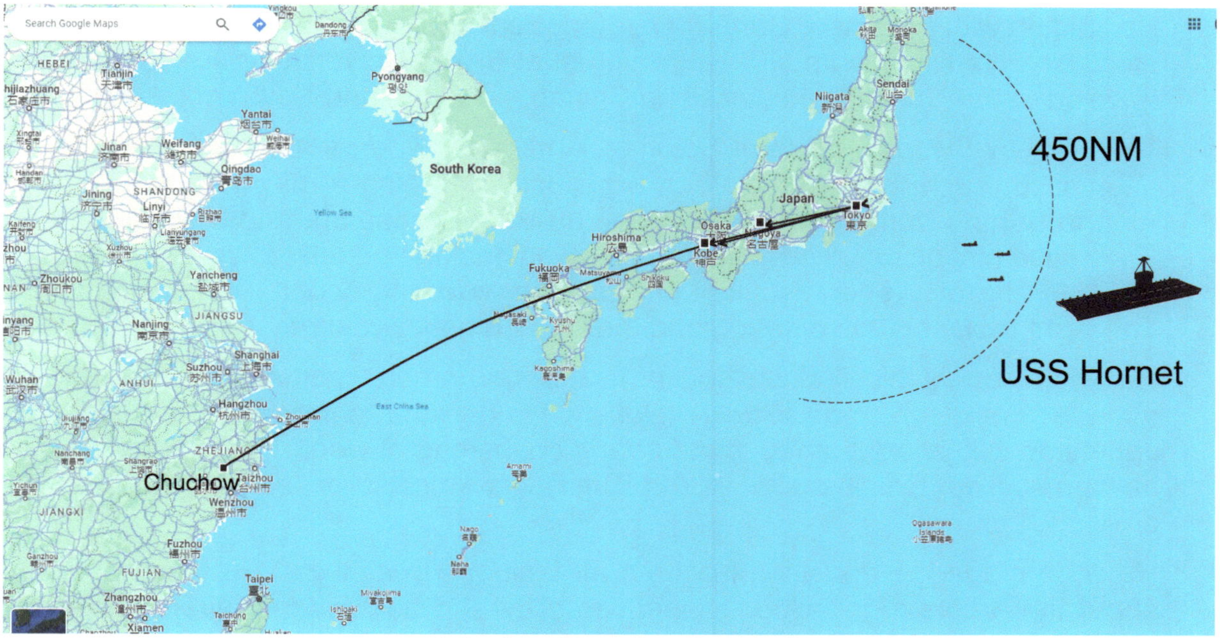

Mission Planning Details[6]

James Doolittle became one of the most important American figures during the Second World War. He received the Congressional Medal of Honor for this achievement.

THE PASSION OF ALBERTO SANTOS DUMONT

Alberto Santos Dumont was a Brazilian pioneer in the history of world aviation for his inventions, including the Dirigible XX, the 14 Bis airplane, and the Demoiselle airplane. He also flew his XIV Bis airplane for 60 meters at two to three meters above the ground at the Champ de Mars, in Paris, on October 23, 1906. The flight was recognized by the Aeroclub de France as the first flight of a heavier-than-air, self-propelled aircraft in Europe.

Born in the town of Palmira, 240 km from Belo Horizonte, on July 20, 1873, Santos Dumont would come to be considered by many as the father of aviation for his achievement in October 1906 in Paris.

Santos Dumont was the son of a wealthy family of coffee growers. His father, Henrique Dumont, worked on the Dom Pedro II railroad and later bought a farm about 20 km from Ribeirão Preto, São Paulo, where he grew coffee. The sixth son in the family, Alberto Santos Dumont was interested from an early age in the locomotives his father had on the property, and in the coffee production machines. Later, encouraged by his parents who gave him part of their fortune during his lifetime, he moved to Paris to study science and engineering. An admirer of Jules Verne's books, he was dedicated to studying and perfecting balloons.

[6] Based on https://www.nationalmuseum.af.mil/Upcoming/Photos/igphoto/2000484315/

He studied with Professor Garcia, who had great knowledge of physics, mechanics, and electricity. He was also a student assistant at Bristol University in England. In 1898 he traveled to the United States, visiting Chicago, Boston, and New York.

The book by Swedish explorer Solomon Andreews, *Andree and his Balloons,* about his attempt to reach the North Pole in a balloon, was also a major influence on his life. Alberto Santos Dumont met the balloon builders Lachambre and Machuron in France. They gave him all the information and instructions he needed to fly a balloon. On March 23, 1898, he made his first flight in a balloon, covering about 100 km with Machuron. The flight took about two hours, and the balloon had a volume of $750m^3$.

He soon bought his first balloon from the Lachambre factory. A small 100 m^3, the manufacture of which he followed closely. He decided to improve the design by using Japanese silk instead of fabric, and bamboo instead of wood. He took to the air in this balloon - called "Bresil" on July 4, 1898, in Paris, which surprised everyone with its small size.

He soon made the transition from hot air balloons to airships. He then began to think about the problem of balloon maneuverability and built his first airship using internal combustion engines, given the heavier weight of electric motors. He built the Number 1 airship, which had an elongated shape, was inflated with hydrogen, and was powered by a gasoline engine. From there, carrying out various experimental flights, he developed airships 2 and 3.

In 1901, French millionaire Henri Deutsch de la Meurthe offered a prize of 100,000 francs to anyone who could fly around the Eiffel Tower from Parc de Saint-Cloud and, without any ground support, return to the same starting point. The journey had to be completed in no more than 30 minutes.

On July 13, 1901, he took off with his airship Number 5 and circled the Eiffel Tower, but due to an engine failure, he lost control and crashed into the trees in Parc Edmond-de-Rothschild.

On his second attempt, aboard airship Number 5, he once again managed to get around the Eiffel Tower, but the airship lost some of its hydrogen and shape, causing the propellers to break the suspension ropes, which forced Santos Dumont to cut the engine. The airship crashed into the roof of the Trocadero hotel, causing a huge explosion. The accident was almost fatal. Santos Dumont was saved because he had tied himself to the rudder of the airship and was rescued by the Paris fire department.

Finally, using his airship Number 6, Santos Dumont managed to complete the journey and landing of the airship in 30 minutes and 30 seconds. As the landing was also to be taken into account, after deliberation, the prize of 129,000 francs (20,000 dollars) was awarded to Santos Dumont. The President of Brazil, Campos Salles, also awarded him a prize of equal value and a gold medal with his effigy and the inscription "For skies never before navigated."

Alberto Santos Dumont taxiing the 14 Bis in Paris

Winning the Deutsche prize was an important milestone in Santos Dumont's life and was also the incentive he needed to continue his work. The result of his flight also reveals the precision of his design and the control he obtained over the speed and trajectory to complete the route in exactly the time available.

The Number 9 airship was probably the most successful in Paris, as it was smaller in size and he used it for everyday tasks, such as shopping and visiting friends. It was akin to what today would be a "flying car.." Airship Number 10, made for 12 passengers, was also remarkable. And the subsequent Numbers 11, 12, 13, and 14 followed. As can be seen, he managed to develop an increasingly sophisticated series of airships.

However, it became clear to him that airships would not be the answer for flights where the weather conditions were very changeable with strong winds, precipitation, and day and night flights. Even with these limitations, airships are still used today to transport cargo and people, especially in remote regions.

It was then that events came along to boost the evolution of aviation, and also the technologies used by Santos Dumont: the Archdeacon Cup and the Prix d'Aeroclub de France. The Archdeacon Cup was established in 1906 by Ernest Archdeacon, an aviation enthusiast and one of the founders of the Aeroclub de France, the oldest aeroclub in the world. The Cup rewarded a flight of at least 25 meters in a heavier-than-air, self-propelled aircraft.

The Aeroclub de France Prize was created in 1904 by the Aeroclub and was awarded to the aviator who managed to fly a distance of at least 100 meters. The prize was worth 1,500 francs (300 dollars).

To win these prizes, Santos Dumont built 14 Bis. This was an aircraft that used a 50 hp gasoline engine and a propeller in the rear position to push the aircraft ("a push propeller"). It used the kite box concept. The box-shaped kite was able to generate lift due to its shape and curvature. 14 Bis was made of silk and bamboo, with aluminum joints. It had a canard geometry with positive dihedral wings to increase lateral stability. The forward position box was movable to provide longitudinal control.

The 14 Bis takes off from Champ de Mars (Musa et al., 2001)

Santos Dumont won the Archdeacon prize for flying 14 Bis on October 23, 1906, a distance of 60 meters between two and three meters high and lasting seven seconds. It was the first officially recorded flight in history in a self-propelled aircraft. It entered aviation history and was recognized worldwide.

Shortly afterward, on December 12, 1906, he flew 220 meters at a height of six to seven meters above the ground for 21 seconds, reaching an average speed of 41 km/h.

With these achievements, Santos Dumont went on to win both the Archdeacon Prize and the Aeroclub de France Prize.

But that same year, 1906, the Wright brothers finally made public their innovations in the development of the Wright Flyer. The control of the aircraft in flight was admirable and far superior to what had been known until then. This led to great efforts being made to develop self-propelled aircraft and concepts similar to the Wright Flyer.

However, Santos Dumont's efforts would not be in vain, and many of his innovations are still used today. His legacy as a builder and aviator continues to influence people of all ages.

Santos Dumont's first balloon, "Le Bresil," July 4, 1898 [7]

[7] Public domain.

THE PERSISTENCE OF FRANK WHITTLE, ROLF DUDLEY-WILLIAMS, AND JAMES TINLING

Frank Whittle was a British man born in 1908 in Earlsdon, Coventry, England. From an early age, he was fortunate to have the opportunity to learn about mechanical work in his father's workshop. His father was a talented engineer and had a workshop with a few lathes and a single-cylinder internal combustion engine.

At the age of fifteen, Frank decided he wanted to join his country's Air Force. So, he took the tests but failed the physical exam due to his 5ft height and chest diameter. He then decided to train intensively for six months to repeat the test and be accepted but learned that it wasn't possible to reverse the opinion. Whittle then applied to another Air Force apprentice school under an assumed name and was accepted.

He started his apprenticeship. He performed very well in science subjects, especially mathematics. He was recognized by the wing commander of the Apprentice School for his work on model aircraft. The commander was so impressed with Whittle's model airplane work that he recommended him to the Royal Air Force (RAF) pilot school.

At pilot school, located at RAF College Cramwell, he began his training on the Avro 504. He became known for his low-flying maneuvers and acrobatics. He graduated as a "well above average" pilot. However, his book also contains some red marks for low flying and aerobatics, and he was disqualified from the best pilot competition.

The course required a final dissertation, and Whittle wrote a paper entitled "Future Advances in Aircraft Design." In this paper, he described that future airplanes would have to be able to fly faster and that for this to happen they would have to fly higher, where the air is less dense. For this to happen, new aero engines would have to be developed because the piston engines of the time lost a lot of efficiency as the altitude increased and the density of the air decreased.

His idea was to use a piston engine to drive a compressor that would throw compressed air and fuel into a combustor to be burned. The gases themselves escaping at high speed would generate the traction needed for flight. This idea was later dismissed by Whittle himself when calculations showed that this engine would have the same weight as a conventional engine.

So, he decided to replace the piston engine with a turbine, similar to turbochargers, and from this came the basic concept of what would later become his aircraft turbine. Turbochargers use a turbine associated with a compressor via a single shaft. The exhaust air is directed to the turbine, which also drives the compressor.

It was later discovered that the compressors of the time worked "stolidly" and that if they had an aerodynamic profile like wings, their efficiency would be dramatically increased.

Whittle presented his idea to the Air Force, which at the time had little knowledge of the subject. The article was then sent for analysis to Mr. A.A. Griffith, who was a renowned engineer in the field of aero engines. But Mr. Griffith found an error in Whittle's calculations and took little notice of his work. Shortly afterward, Whittle received a letter from the RAF denying support for his project.

His Air Force friend, Pat Johnson, was the one who encouraged him to apply for a patent, which he obtained in 1930.

However, in 1935 Frank Whittle could not afford to renew his patent and his work became public. Available to anyone, it quickly became known in several countries, including Germany at the very time Hitler was building his new Luftwaffe.

After graduating as a pilot, Whittle was appointed to the flight instructor course. He turned out to be an excellent instructor. As the Air Force required officers to choose a specialization, Frank Whittle was nominated for engineering. He excelled academically, especially in mathematics, and was appointed to study at the famous Peterhouse School, the first Cambridge College School, in 1934.
At Cambridge, he had the opportunity to deepen his knowledge of engines and refine his jet engine project.

It was then that an event changed the story of this then-little-known RAF officer. He received a letter from a former colleague at Cranwell Base, Rolf Dudley Williams. In the letter, Williams described his interest in Whittle's project and the possibility of setting up a company to take the idea forward and make it operational. He went to Cambridge to talk to Whittle, along with another person who became very important in the project, James Tinling. Together, they convinced Whittle to set up a company and obtain funds to build the engine. Thus, was born the future Power Jets company, which was officially created in March 1936. Power Jets was a partnership between these three men and the investment bank, Messrs. O. T. Falk and Partners. The initial capital in the new company was 2,000 pounds, and British Thomson-Houston (BTH) was awarded a contract to build the engines in June 1936.

First Engine Design Criteria

This engine had a single-stage centrifugal compressor and a single-stage turbine. According to Frank Whittle[8], the objectives of the project were:

- Get at least 3,000 shp.
- Obtain a compression ratio of 4:1, which was much higher than had been obtained at the time (around 2 ½).
- Obtain a much higher combustion temperature than was previously possible.
- Do this with high efficiency.
- Equip an aircraft weighing around 2,000 lb.
- Fly at a maximum speed of 500 mph.
- Fly at a maximum altitude of 70,000 ft.

The thermodynamic design of the engine was similar to the figure below.

[8] WHITTLE F. The early history of the Whittle jet propulsion gas turbine, Institution of Mechanical Engineers, James Clayton Memorial Lecture, London, 1945.

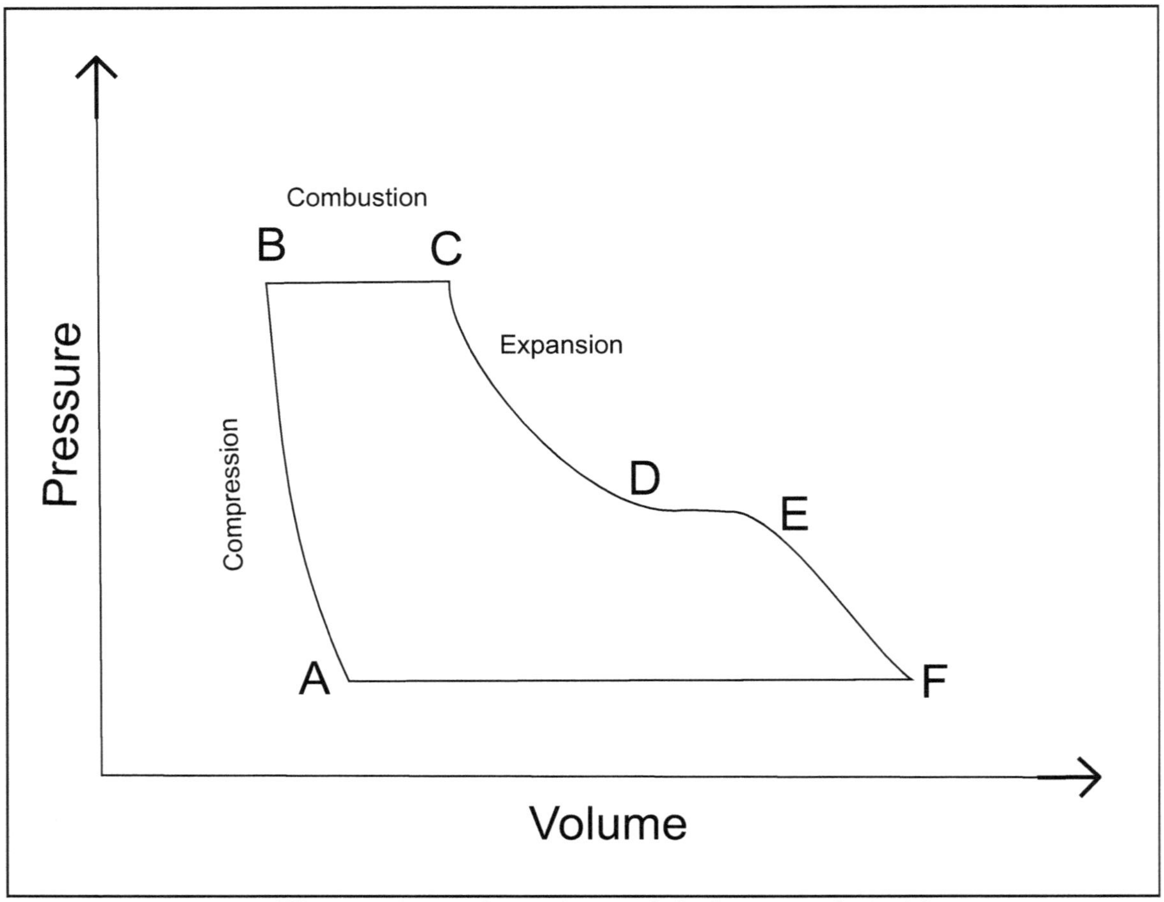

Thermodynamic cycle model of a generic turbine

Whittle said they had confidence in the design of the compressor and turbine but had many doubts about combustion. That is why they looked for combustion experience in British refineries. Most of them found the combustion design criteria very difficult until Mr. Laidlaw, from Laidlaw, Drew and Company, decided to take part in the project.

Tests with the first engine

A little over a year later, in April 1937, the first engine, the Model U, was ready for testing. These first tests showed that the combustion problem had not been solved and that the efficiency of the compressor was well below expectations. A new engine was therefore ordered from the British Thomson-Houston Company.

The Model U engine start sequence was as follows:
- Accelerate the engine to 1,000 rpm with the starter motor.
- Operate the pilot flame using a spark plug and a hand-held magnet.
- Accelerate the engine to 3,000 rpm and open the main fuel control slowly.
- The engine then accelerated under its own power until it stabilized.

However, after the last stage of starting, the engine accelerated out of control to around 50% of the maximum design speed, much to the amazement of those involved. This out-of-control acceleration posed a great risk of explosion. This phenomenon was later understood and attributed to the engine's fuel system, which injected an excessive amount of fuel into the combustion chamber when the fuel pump was switched on. This was because, when the burner nozzle was not completely full, the fuel pushed the injector nozzle to the fully open position. Other out-of-control accelerations were attributed to the loss of burner spring temperature due to overheating.

Fortunately, there were no problems when the ignition failed due to poor contact between the electrodes, or problems with the fuel line.

Overheating in the combustion chamber was also a major problem with U engines, which led to major design changes. One of these was to move the injection to a more downstream position. But this made combustion very inefficient, preventing the engine from accelerating. Many modifications were made to the combustion chambers and some improvements were achieved. One of the breakthroughs was to run the engine for 10 minutes at 11,000 rpm. However, the compressor failed when it remained at 12,000 rpm for four minutes. They also ended up changing the fuel from diesel to kerosene to improve combustion.

Another shortcoming was that the compressor could not deliver the required compression. It was then decided to fit diffuser blades in an attempt to improve the compressor's efficiency, which resulted in a speed of 13,000 rpm but with a compression ratio that was still lower than desired.

It was also noted that the flow between the compressor and the combustion chamber appeared to be reversed. This was later effectively proven when flames were seen coming through a hole drilled in the 90-degree junction (elbow). Several modifications were made to the compressor and combustion chamber to try to solve this problem, and they eventually opted to use a new set of diffuser blades when they reached a speed of 13,600 rpm.

During this first phase of testing, it was decided by the chief engineer of British Thomson-Houston that the maximum speed could not exceed 12,000 rpm inside the factory for safety reasons. Tests with the first engine were, therefore, limited to this rpm. Afterward, the injection returned to its original position in the combustion chamber, but this time with a burner protected from overheating by a fuel strip. And for the first time, no part of the assembly reached temperatures high enough to achieve an incandescent glow.

At the end of the testing campaign for the first engine, some conclusions could be drawn:
- The compression ratio was lower than necessary.
- Combustion needed to be improved.
- There was excessive preheating of the air taken from the rear intake due to the shape and temperature of the combustion chamber.
- Excessive friction loss due to the shape of the turbine exhaust.
- Despite the difficulties, the engine seemed viable.
- The tests would have to be carried out elsewhere due to the dangers involved (mainly of explosion).

In 1937, the UK Ministry of Air Defense took a greater interest in the project, on the recommendation of the Aeronautical Research Committee, and signed a contract with Power Jets. The project was classified as confidential, which made it much more difficult to obtain private funding. But this proved necessary, especially after a German magazine called *Flugsport* published several drawings of Whittle's patent in 1939.

Tests with the second engine

The design of the second engine aimed to have a single straight combustion chamber and an improvement in efficiency in the compressor diffuser system. The compressor was to be symmetrical about its axis. The turbine's airflow was reversed. After several combustion tests, and with the help of Laidlaw, the injector nozzle was modified to have a helical shape with ten nozzles to vaporize the fuel and inject fuel against the airflow.

They restarted testing when the second engine was ready in April 1938. The second engine had significant design differences. One of them resulted from the hypothesis of speed at the turbine outlet. For Whittle, the exit speed should follow the principle of conservation of angular momentum. Many British Thomson-Houston engineers, however, believed that the exit velocity would be that of jets with almost constant velocity and pressure. This explains why there were so many discussions about the design of the turbine and why the torsion of the turbine blades was less than necessary, compensating only for the difference in speed along the blade. In the second engine, this error was corrected, and the turbine blades had approximately double the twist.

The design was modified to take into account a resulting traction of 1,240 lb., as opposed to the 2,000 lb. at the beginning. Measured traction, however, was only 480 lb. The compressor efficiency was revised to 70%, the same as the turbine.

Right at the start of the tests, after about two weeks, the turbine blades failed, severely damaging the engine. Only nine runs were made with this engine. In the first eight rounds, it did not go above 8,500 rpm, and in the last, it reached 13,000 rpm for 20 minutes, when the turbine failure occurred.

It was later observed that the turbine's metal casing support was not strong enough to withstand the pressure differences between the compressor and the turbine, and this caused the turbine blades to rub against the casing wall.

In addition, the combustion control problem had not been solved.

Tests with the third engine

It was then decided to rebuild the engine. While the engines were being rebuilt, they carried out new tests to improve combustion control. Tests on the third engine began in October 1938 and continued until February 1941 when a fault in the turbine disk caused irreparable damage. This third model, however, made great contributions to future projects.

In 1939, given the progress demonstrated by the project, and the prospect of producing a viable engine to equip an aircraft, the British Minister of Air Defense signed contracts with Power Jets to produce an aircraft engine and with Gloster for the design and production of the aircraft that would be equipped with the engine. This first engine, produced for use during flight tests, was called the W1.

During the production of the W1 engine, however, some important parts had production problems and would not be used for flight. For this reason, it was decided to use this engine for new bench tests. This engine was called the W1X and was delivered for testing in November 1940. Some of the modifications tested on the W1X were incorporated into the W1 engine.

When the W1 engine was delivered, it was decided to put it through a 25-hour running program to clear it for flight. At the same time, Gloster had completed production of the E28 aircraft, so it was decided to install the W1X engine to cab the E28 aircraft in April 1941.

In May 1941, the engine was cleared for a ten-hour flight, and test flights began. The tests lasted about ten days, and the test pilot was Flight Lieutenant P.E.G. Sayer.

Gloster E28/39 aircraft[9]

At the beginning of 1940, the Minister of Air Defense had authorized the production of a new engine, which was named W2. He also signed a contract with Gloster to produce a twin-engine jet aircraft for air interception missions. This aircraft would be called the F9/40, the future Gloster Meteor jet, the first jet aircraft to become operational in the British Air Force. The Minister for Air Defense also signed contracts with British Thomson-Houston for engine production and made PowerJets an aero engine design and research organization, giving all the necessary support to the other companies involved in engine manufacturing.

[9] https://en.wikipedia.org/wiki/Gloster_E.28/39

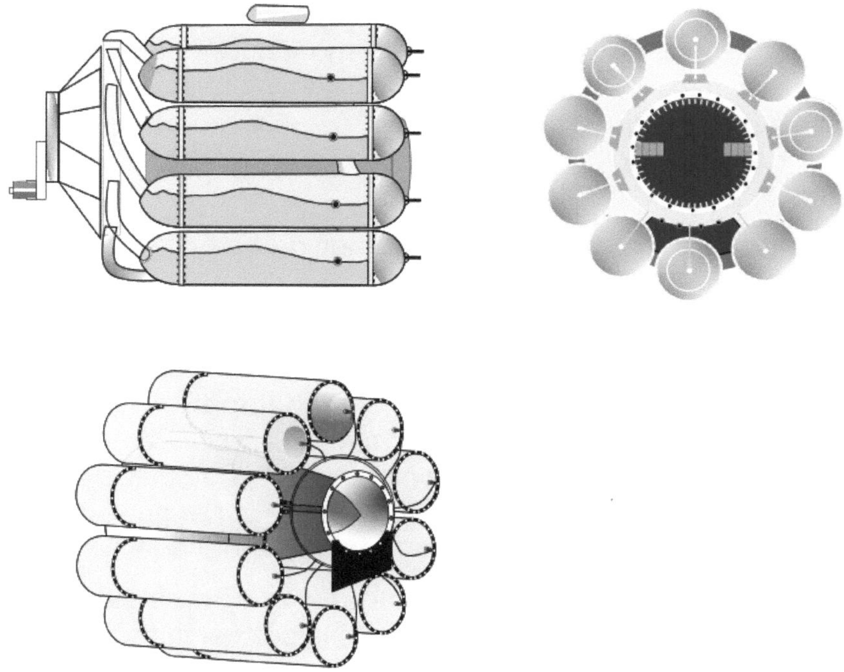

W.1X Experimental Engine[10]

[10] Based on the article Whittle. F. "The early history of the Whittle jet propulsion gas turbine"

THE WRIGHT BROTHERS' VISION

The Wright brothers, sons of a bishop of the Church of the United Brethren in Christ, are classic examples of teamwork and a clear vision. Over about ten years of work, the "bishop's boys" managed to evolve their invention from a glider to a motorized aircraft capable of carrying two passengers in a controlled flight. They were the first human beings to achieve this feat.

Milton Wright and Susan Catherine Koerner had seven children. They lived mostly in Dayton, Ohio, where Wilbur was born, but due to his work in the Church, Milton Wright moved ten times when his children were growing up.

When he was about six years old, Wilbur had an accident during a field hockey game in which he was hit in the face and lost his front teeth. This greatly affected the boy's personality, making him serious and introverted, which later also changed his plans to go to Yale. He ended up only finishing high school. Due to the constant moves Reverend Milton had to make, Wilbur finished high school but ended up not being able to obtain his diploma, which was awarded to him by the school posthumously.

His younger brother, Orville, became the owner of a small printing company, printing newspapers and periodicals. Wilbur joined his brother in the venture. Later, the brothers joined forces in a new venture, this time a bicycle shop, taking advantage of the boom in bicycles at the end of the 19th century, known as the bicycle craze." With the profits from the bicycle shop, they were able to begin their first experiments with their new passion, aviation.

When they learned about the experiments of other aviation pioneers, especially the German, Otto Lilienthal, who in the 1890s made hundreds of experimental flights with his gliders, they began to study the problem of heavier-than-air flight. In 1889, Lilienthal published a book that would inspire many other pioneers *The Flight of Birds as a Basis for Aviation*, in which he analyzed, among other things, the wings of storks. He died in an accident involving one of his gliders in 1896 when he was just 48 years old. But that did not dampen the brothers' interest in his project.

Using Lilienthal's aerodynamic data and Octave Chanute's glider design as a reference, they built their first aircraft, the "Wright Flyer" glider. However, unlike most aviation pioneers, they decided to keep their design a secret, so that they could avoid reporters and later, perhaps, take credit for their inventions through patents.

In the summer of 1899, they experimented with a glider model with lateral control by wing warping. This glider was a biplane with a wingspan of around five feet and a chord of one foot. By twisting the wing, using a system of rods, the wing gained lift, making the glider turn to the other side.

In the summer of 1900, they planned to begin flight trials of their first full-scale glider. To have the ideal conditions for the experiments, they decided to take their aircraft to North Carolina, to a place with enough wind to facilitate gliding. The site was near Kitty Hawk, called Kill Devils Hill, and is where Wright Petterson Air Force Base is today. There was a high enough hill and that is where they started their experiments.

Historic photo: Wilbur Wright after landing his 1900 glider

The 1900 glider had a wingspan of only 5.3 m and a wing curvature of 1:22. It had a canard configuration in which the depth control was positioned in front of the wing. It had no directional control, today called a rudder. The glider initially flew without a pilot, like a kite, and was later piloted by Wilbur Wright. He made twelve flights in this glider, which was left at Kitty Hawk, where it was eventually destroyed by strong winds. One of the great lessons from this glider was that the "horizontal rudder," now known as the elevator, allowed longitudinal control of the glider and that lateral control by wing warping was possible.

Later, Orville wrote that they used the canard configuration for many years because, when it stalled, the glider fell like a leaf, unlike the conventional configuration, where it usually went into a tailspin, which often had fatal consequences. An intriguing fact was that the lift they calculated was much lower than the predictions obtained using the aerodynamic tables of Otto Lilienthal and James Langley and John Smeaton's equation.

The 1901 glider had a larger wing with a 6.7m wingspan and a wing curvature increased to 1:12. It also had a fixed rudder. This glider taught the Wrights important lessons. For the first time, they found stall, which made the glider fall horizontally to the ground. They once again noticed that lift was much less than they had calculated using John Smeaton's formula - only about a third of what was predicted. In addition, drag was also much lower. They therefore began to question the Smeaton coefficient (there was still no concept of air density). Through experiments carried out in their wind tunnel, they proposed that the Smeaton coefficient be corrected to 0.0033 instead of 0.005. For the first time, they noticed that the wing that lifted during a turn tended to fall back, generating yaw. This was the phenomenon known today as adverse yaw.

The 1902 glider, eventually led to the design of the Wright Flyer. It was in the 1902 glider that they replaced the two-part fixed rudder with a movable rudder. This greatly increased directional stability and reduced the "adverse yaw" they identified in the 1901 glider.

The Wrights spent from 1900 to 1902 testing gliders at Kitty Hawk. In all, they built three gliders. The gliders of 1900 and 1901 brought important discoveries, and the main one was that the wing was not generating the lift they had hoped for. They solved this problem by increasing the wingspan and wing area of the gliders. They also solved the important problem of directional stability by using a directional rudder. Furthermore, they identified a problem in Smeaton's constant and in the aerodynamic data of other researchers and pioneers, such as Otto Lilienthal and James Langley.

Historic photo: Test of one of the Wright Brothers' gliders

The Wright Flyer was a canard-style biplane, with the canard at the front of the aircraft and a single 12-hp gasoline engine. It had two propellers, both connected to the engine, which turned in the opposite direction. Lateral control was achieved by twisting the wing, or "wing warping" It also had a single rudder. Wilbur Wright made the first sustained flight of a heavier-than-air aircraft on December 17, 1903, with the Wright Flyer. But despite his best efforts, it was very difficult to fly and he only made four flights. It was damaged on the last landing and strong winds completely destroyed it later.

The brothers used spruce wood for straight structural items, such as spars, and ash wood for curved wing components, such as ribs. The engine had four cylinders and was designed by their employee, Charles Taylor, to be lightweight, as they could not find a suitable car engine. It was powered by a one-gallon (3.8-liter) tank.

Historic photo: The first flight of the Wright Flyer

The Wright brothers filed a patent application for the Wright Flyer in 1908.

THE MAN WHO COULD SEE THE WIND, AND HIS TEST PILOT

Clarence Leonard "Kelly" Johnson was born on February 27, 1910, and died on December 21, 1990. He was the son of Swedish immigrants. This boy from Ishpeming, a remote town in the state of Michigan, was to become one of the greatest aircraft designers in history, having developed 44 different aircraft throughout his career, among them icons such as the U-2 reconnaissance plane, the fastest airplane in history, the SR-71 Blackbird, and the first American jet fighter, the P-80 "Shooting Star". He was the only man to win the famed Collier Prize twice.

He started at the age of 13, winning a school prize for designing his first aircraft, which he called the Merlin fighter plane. Before he had seen an airplane in real life.

When still at school, some boys started calling him Clara because of his name, Clarence. One day, he decided he had had enough of it and settled the matter by getting into a fight with one of the bullies. The boys at school then decided that he should have a new nickname, and as there was a famous Irish boxer called Kelly, that is what they called him.

He made his first flight in a biplane for five dollars, but it ended badly when the engine lost power at an altitude of 700ft. Later, with the money he earned working in construction and at the Buick car factory, he enrolled in a flying school. The instructor, a pilot of limited financial means but enthusiastic about aviation, told the young man to spend his money on college.

Johnson started at Michigan College in 1929, paying his way by washing dishes in fraternity houses. It was then that he started as an assistant in the Aeronautical Engineering Department, working in the wind tunnel. The college allowed him to use the tunnel to provide services when it was not in use. He started charging $35 an hour for the wind tunnel, which he used to make aerodynamic improvements to the tunnel, also helping design a model car for the Studebaker factory.

When he graduated in 1932, he tried to enlist in the Army Air Corps but was rejected and decided to return to college to do his master's degree. Among other activities, he helped design Indianapolis race cars using the university's wind tunnel.

The Electra 10

In 1933, Johnson got a job at Lockheed as a tool designer. At the time, Lockheed was in a very unstable situation, having emerged from bankruptcy, and was betting heavily on its new twin-engined transport plane, the Electra. Johnson convinced chief engineer Hall Hibbard that the airplane would be unstable directionally, especially if one of the engines failed. Hibbard gave Johnson a model of the Electra to test in the wind tunnel and see if he could improve the design. After 72 rounds in the tunnel, Johnson suggested redesigning the flap system and introducing a twin tail, which would become the aircraft's hallmark and help make it one of the most successful aircraft of its time. Johnson was 23 years old.

When he returned to Lockheed, he was promoted to aeronautical engineer and began a career as a Flight Test engineer on the Electra 10 program. He also worked as a load, weight, and balance engineer, and aerodynamicist. He worked on the various Electra models, attracting attention both inside and outside the company. In 1937 he was awarded the Sperry Prize for his unusual achievements in the field of aeronautics. In total, he flew around 2,300 hours as a Flight Test engineer.

Johnson and the SR-71 Aircraft[11]

The XP-38 Lightning

When the United States was preparing to enter the Second World War, the military set up a competition to develop a new fighter jet.

To take part in the competition, Lockheed created a team of engineers to work in secret, a concept that would later be called "Skunk Works," and which would be led by Hall Hibbard and Kelly Johnson. Lockheed won the competition with an aircraft designed by Johnson, the XP-38. It was an aircraft with different characteristics, with two engines and two fuselages, the cockpit sitting between the two fuselages, and a double tail. The P-38 was the United States' first long-range fighter, and each of its engines was equipped with superchargers, which greatly improved its performance at high altitudes.

[11] Courtesy from Lockheed Martin Company.

There was also an important evolution in the construction of the aircraft: its fuselage had aluminum panels joined by countersunk rivets, which greatly contributed to reducing drag and weight. As a result, the P-38 was the first fighter to exceed 400 mph.

At first, it had problems diving at high speed, when the aircraft tended to steepen its dive, slipping out of the pilot's control to the point of breaking up in the air.

Lockheed P-38 Lightning[12]

Returning to wind tunnel tests to analyze the problem at high speed, Johnson discovered the characteristic known as "Mach Tuck," which occurred due to the formation of shock waves over the wings. After some aerodynamic modifications proposed by Johnson, the P-38 passed the Army Air Corps tests. It was to be the only fighter aircraft mass-produced from the beginning to the end of the Second World War. Johnson was only 27 years old at the time.

The P-38 roll ratio was low compared to other fighters of the time and was later improved with the introduction of hydraulically actuated ailerons.

After the outbreak of the Second World War, P-38s were produced by the thousands (more than 10,000 were produced during the war) and Johnson followed the production of these planes closely, becoming a master at finding production and design improvements to make the production of planes easier and cheaper, creating more efficient processes.

---

[12] Public domain.

The Hudson Bomber

In 1938, with the growing threat of war, Great Britain's Royal Air Force sent a commission to evaluate the possible purchase of a maritime patrol and anti-submarine aircraft. The commission visited several manufacturers, but Lockheed was not initially on the list. When the British included Lockheed at the last minute, there were only five days left. Johnson had to draw up a proposal for the British in just five days. The proposal was to transform the Electra to meet British requirements. In addition, they built a full-scale model of the new plane. This shows one of the characteristics that would make him famous - the ability to create completely new aircraft from existing designs. The commission was so impressed with the proposal that it called Lockheed to come to England to present it to the High Command. When they arrived, however, the requirements had changed significantly. Johnson locked himself in a hotel room for three days straight and came out with a proposal modified to meet the new requirements. This once again impressed the British, who decided to sign a 200-plane contract with Lockheed Martin to produce the aircraft that would become known as the Hudson Bomber. On his return to the United States, Lockheed promoted Johnson to the position of chief engineer.

Johnson believed in practicality, concise reports (no longer than 20 pages), and small, well-prepared teams focused on a task. One of the sayings that expresses his way of working well was: "Be quick, talk little, and deliver on time."

In 1943, Lockheed made him head of the Advanced Projects Section. This ended up being called the "Skunk Works," originating from a comic book that had a mysterious and powerful elixir called "Kickapoo Joy Juice," produced in a factory called the "Skunk Works." This elixir was the result of a lot of strange ingredients such as prawns and old sneakers being mixed together. Something similar happened in this section at Lockheed, as no one knew for sure what Kelly Johnson was building there.

The Hudson Maritime Patrol Aircraft[13]

The P-80 Shooting Star

The Skunk Works' first project was the P-80 Shooting Star, signed off in 1943. The US Air Force and the Allies had information indicating that the Germans were well ahead in the development of jet aircraft, which was a great risk for the operations. The big boost was Intelligence's discovery of the German Me-262 jet fighter. Propeller-driven fighters were no match for jet aircraft that flew faster and higher. Almost simultaneously, the British had developed their first jet engine, the Gloster E28/39 aircraft, thanks to the efforts of Frank Whittle and the Power Jets company. The engine for the P-80 would be the Halford H-1 Goblin, an evolution of Frank Whittle's W1 and the first jet engine to be certified in England.

The P-80 contract had an absurdly short deadline of just 180 days for delivery of the first aircraft. Lockheed managed to make the first flight of the P-80 37 days ahead of schedule. In just five months, Lockheed had designed and built the first American jet fighter.

It was a highly maneuverable aircraft, and faster than all the others available to the US Air Force at the time. It had non-swept wings because at the time the advantages of this type of design for reaching higher speeds were not yet known, and its fuselage was made entirely of aluminum. Despite this, the P-80 was the first North American aircraft to reach 500 mph.

[13] Public domain.

After the first flight, test pilot Anthony Lavier found a tendency for "porpoising" due to errors in calculating the center of gravity. A pressurization valve caused the engine to bleed air at 325°C to enter the cabin. In addition, problems with the flaps led to the aircraft rolling violently. Lavier was able to bring the aircraft back safely thanks to his exceptional piloting skills. It was the beginning of a long-term partnership between Johnson and Lavier. Besides the P-80, Lavier would do the first flights of many important aircraft like the F-94 Starfire, the T-33-1, the XF-104 Starfighter and the U-2 reconnaissance aircraft.

The war ended before the P-80 could be used in combat. But the P-80 was used extensively in the Korean War when it was replaced by the transonic F-86 Sabre. The T-33 jet trainer, a two-seat version of the P-80, was used by the Air Force until the mid-1980s, and the Navy's NT-33 was retired in 1997, with an extensive service life of more than 50 years. Some 5,691 T-33s of various models were made by Lockheed.

The Constellation

The Constellation was a four-engined aircraft designed by Lockheed that was used for both civilian and military transportation. The requirements for this airplane were created by billionaire aviator Howard Hughes, the majority shareholder in Transcontinental Western Airlines (TWA). Hughes' requirement was for a 40-seat transcontinental aircraft with a range of 3,500 nautical miles.

The L-049 Constellation turned out to be the first pressurized aircraft to be widely used for civil transport across the globe. It had a particular design with a triple tail and a dolphin-shaped fuselage. The fact that it was pressurized allowed it to fly higher and, therefore, above the altitude range where weather conditions are most critical, so increasing flight safety.

It also had a de-ice system on the leading edges of the wings and empennage, as well as hydraulically assisted flight controls. It had an operational ceiling of 24,000 ft and a top speed of 600 km/h (which was, by way of comparison, faster than the Japanese Zero fighters).

The U-2 Dragon Fly

In the years following the Second World War and with the emergence of the Cold War between the United States and the Soviet Union, Kelly Johnson's Skunk Works would play a major role, becoming the most important aircraft design office in the world.

The 1950s were to become a decade of great technological progress but driven by a constant nuclear threat.

With the intelligence agency's need to obtain more information about the Soviet ballistic and long-range bomber program, the U-2 was to play an important role.

In the design competition, the best one capable of flying at 70,000ft, where the existing Soviet Mig-17 fighters and anti-aircraft artillery could not reach, would win. Although Lockheed was not invited to participate, Johnson submitted a proposal for an aircraft with a fuselage similar to the F-104, but with a high aspect ratio and wingspan. Despite losing the competition to the Bell X-16, Kelly Johnson did not give up. After making some improvements to the design, he was awarded a $22 million contract to produce 20 aircraft and make the first flight in just eight months.

The CIA director at the time was very pleased with the way Skunk Works developed aircraft, with few people involved, experienced, qualified teams, and aggressive targets. Only a small group of engineers and technicians had access to the project, and this was what Johnson wanted, as it avoided the need for too many explanations.

To reach high altitudes they had to ruthlessly reduce structural components and weight. It had glider-like wings with a wingspan of 24.3 meters and weighed only 190 N/m2. The engine would be a version of the J57 modified for high altitudes, with minimum production tolerances and a thrust of 10,500 lbf.

On August 1, 1955, Tony Lavier climbed into the cockpit of the U-2 for taxi testing. On the third run, he noticed that the aircraft was flying about three meters above the ground. When he landed, the tires burst, and the brakes caught fire and failed. Lavier had no intention of flying at all, as he would later comment.

Four days later they would complete their first flight and Lavier commented that the aircraft was soaring through the sky like "a homesick angel.."

Although announced as a research aircraft for NASA, the U-2's mission was to fly above the altitudes attainable by air defense, carrying 350 kg of cameras and other reconnaissance equipment. It went into operation in July 1956, just 11 months after its first flight. And it provided the first data on the Soviet Union's real intercontinental ballistic missile capacity.

The U-2 flew for about four years, collecting data, virtually unscathed until pilot Francis Gary Powers was shot down flying a U-2 over the Soviet Union in 1960. But that did not stop it from continuing to fly reconnaissance missions in various parts of the world, such as during the Cuban crisis in 1962, as well as many high-altitude research missions.

Lockheed returned two million dollars of the original $22 million contract that had been saved during the project. And it built five new planes from parts that were left over.

With some improvements, the United States Air Force continues to operate the U-2 to this day. It is still the highest-flying single-engine aircraft in the world.

The F-104 Starfighter

In 1954, when test pilot Tony Lavier saw the F-104 for the first time, he reportedly asked where the wings were. Born out of design requirements from Kelly Johnson himself, the F-104 was designed to fly at Mach 2 and an altitude of 60,000 ft. When they showed the proposal to the Air Force, the F-104 was immediately well received. In February of that same year, Tony Lavier made the first flight of the F-104 Starfighter from the dry lake at Edwards Air Force Base, in California.

Even without the much more powerful J79 production engines, the F-104 was then the fastest fighter in the world.

Johnson was obsessed with efficiency, simplicity, and cost reduction. As the F-104 project progressed, he created a team of engineers with the task of finding ways to simplify and make production more efficient. This resulted in millions of dollars saved for customers over the years.

The F-104 would become the first aircraft to break the altitude and speed records together, reaching 91,000 ft and 1,400 mph in 1958. Later that year it would break three climb performance records. And in 1959 it would become the first aircraft to climb above 100,000 ft under its own power, when Captain Joe Jordan zoomed to 103,000 ft with his F-104.

These milestones earned Johnson the Collier Prize in 1958.

Kelly Johnson receiving the Collier Trophy from President H. Nixon[14]

[14] Permission for use and reproduce granted by the Archives Division of the National Air and Space Museum, Smithsonian Institution.

THE X-7

A huge amount of German aerospace research data became available for US academic research at the end of the Second World War. There were research opportunities in all areas of aeronautical knowledge. Another important factor was the availability of financial support for research in the post-war period, which helped open up unprecedented possibilities.

The X-7 was the result of a request from the Air Force to design a ramjet engine test vehicle that could reach at least Mach 3. As a result, it was designed to fly at 93,000 ft and reach Mach 4. The aircraft was designated X-7 in 1951, and the project would last until July 1960.

Despite the failure of its first launch, the X-7 made a total of 130 flights, collecting data for a series of experiments in the supersonic regime, missile guidance, and the development of the ramjet engine. The X-7's structure and wings were made of steel and the fuselage was made of nickel alloy, when titanium, which would be widely used in future projects such as the A-12, was not yet used. However, it was very difficult to deal with its extremely toxic fuel.

It was docked under one of the wings of a B-29 or B-50 to be released and then powered by its engine. The X-7's engine was designed to be used for a short period, but enough to take it close to Mach 4. It was recovered by deploying a parachute to make a descent with a low sink rate. At the end of the descent, its long nose buried itself in the ground, preventing damage to the fuselage during landing.

The X-7A was used to test rocket engines and the X-7B was used as a platform for guidance and control tests. The fuel used by the X-7 was very toxic and carcinogenic, as well as being hyperbolic. Hyperbolic fuels do not need igniters, as their components ignite when mixed.

The advances made with the X-7 formed the basis for the AQM-60 Kingfisher program, which was a high-speed target drone that surpassed the capabilities of all the ground-to-air attack systems of the time.

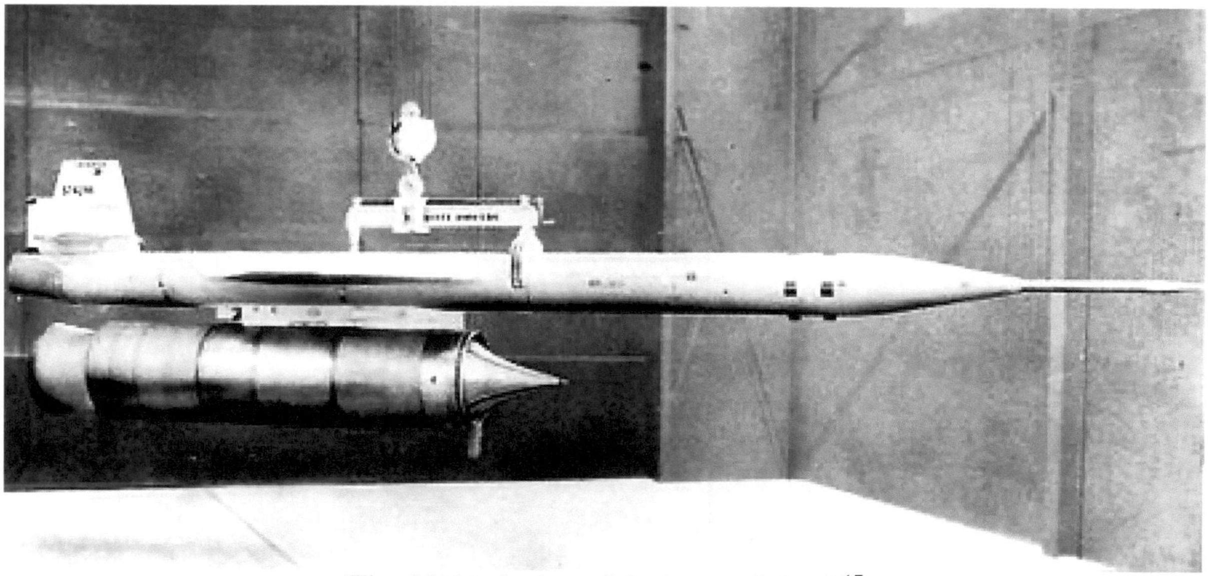

The X-7A-1 aircraft being weighed.[15]

The X-7 weighed 3,600 kg, had a wingspan of 3.7 m, a diameter of 0.51 m, and a length of 10 m. The first version, called the X-7A-1, used a single solid propellant rocket and produced 467 kN for four seconds. In 1954 it underwent significant changes, giving rise to the X-7A-3, which used two solid propellant rockets under the wings, each producing around 222 kN. The way it was attached to the transport aircraft also changed. Instead of a heavy, complicated support system under a wing, it was attached to the fuselage by a much simpler and lighter system.

[15] Public domain.

X-7 recovery[16]

THE XFV-1

It was Lockheed's first program with the Navy. It was an aircraft designed to take off and land vertically on Navy ship decks. The aircraft had a weight and cost within the design standards, but the project failed mainly because of the engine, and because when landing vertically the pilots could not judge the height of the aircraft.

By way of example, the XFV-1's XT-40A-6 engine had the equivalent of 5850 shp. The powerplant had two counter-rotating propellers.

On December 23, 1953, during a high-speed taxiing test, the XFV-1 inadvertently left the ground. Test pilot Herman "Fish" Salmon managed to land the aircraft straight after a few seconds. Despite these problems, the XFV-1 was able to successfully transition from horizontal to vertical flight and still maintain gliding flight in a vertical position.

---

[16] Public domain.

The XFV-1 program suffered constant delays in the delivery of parts, which greatly reduced the ability to keep the pilots operational. Eventually, due to these problems, the program was terminated.

XFV-1 aircraft[17]

[17] Public domain.

The A12 and the SR-71 Blackbird

Two years before a U-2 was shot down in the skies over Russia, Lockheed, on the basis of a contract with the CIA, decided to build an aircraft that would fly higher than the U-2, and four times faster.

The A-12 won the GUSTO project competition between the Convair "Fish" and the Convair "Kingfish" The Kingfish had similar performance but better stealth characteristics, which gave it the edge. Lockheed therefore changed the design to have two inward-sloping vertical empennages and included several parts made of composite material.

It was developed and operated by the CIA's then-named Project Oxcart.

Propelled by a liquid hydrogen engine capable of flying at 90,000 feet and 3.5 times the speed of sound, it was an aircraft like no other. A challenge that required completely new constructive, aerodynamic, and material solutions.

The first studies in 1956 led to the CL-400, an aircraft that would be propelled by liquid hydrogen, which proved to be feasible, but which did not please Kelly Johnson because of its range. He recommended canceling the project in 1957. As an alternative to liquid hydrogen, which has a very low volumetric density, he suggested jet fuel, which would result in a loss of maximum altitude, but great advantages in terms of the size and range of the aircraft.

Johnson received approval to start the project in April 1958 and began a series of design studies.

Some of these studies aimed to reach altitudes of up to 150,000 feet using a combination of rockets, ramjets, and turbines. Some had multi-stage vehicles and others were lifted by balloons.

For the first time, Johnson was faced with requirements that demanded that the aircraft have a low radar reflection, known as *stealth*.

The project was completed in just 16 months and presented to the CIA in August 1959, when it was approved. In April 1962, the A12 was rolled out. It had the appearance of a conventional aircraft, but everything used had to be created especially for it. More than 90% of the airframe was made of titanium to withstand the heat generated at Mach 3.5, around 550°C. No one had ever tried to produce such a large quantity of titanium in the shape and purity required for this project. This required the development of new machining and forging processes. In addition, to obtain the huge quantities of titanium, the CIA set up fake companies to buy titanium around the world. One of the biggest suppliers of titanium was the Soviet Union itself.

When a thin titanium panel wrinkled during a heating test, Johnson had the idea of making corrugated titanium panels. When heated, these panels only got deeper and when cooled returned to their original shape. A very simple solution to a complex problem.

Actively participating in the production of the aircraft, observing, and asking questions was the way Johnson believed the best results would be achieved. The only computer they used was a 25 cm slide rule.

On April 25, 1962, preparations for the A-12's taxiing and take-off were completed. However, during the high-speed runs there was a problem with directional control and the test pilot, Lew Schalk, had to apply the right rudder pedal and full throttle to control the aircraft. The aircraft lifted off the ground and began to oscillate when rolling at low altitude. Schalk was lucky to get the airplane back on the ground.

The next day the problem had been solved and the A-12's first flight was completed with the landing gear locked down. A few days later, Johnson brought a committee of VIPs to watch the first official flight of the A-12. From then on, the project progressed without any major problems.

But it was not until the Pratt Whitney J-75 engines were replaced by Pratt Whitney J-58 production engines, with no less than 32,500 pounds of thrust, that the A-12 showed its real potential. It operated as a normal jet engine with afterburners up to around 1,600 mph when it then switched to an engine bleed cycle in which the air intake produced most of the thrust at Mach 3.

Aircraft A-12[18]

In May 1963, the program suffered its first loss. The aircraft lost control in flight and pilot Kenneth Collins ejected near Wendover, Utah. Collins was rescued by local police. The CIA announced that there had been a crash of a Republic F-105 "Thunderchief" aircraft

[18] Public domain.

and witnesses to the accident were dissuaded from commenting on the matter with the payment of $25,000. Local farmers were told that the aircraft was carrying nuclear weapons to keep them away from the site. The project repeatedly used cash payments to avoid questions about the operation.

In 1964, two A-12s were modified for a different mission - to carry an advanced drone. This was the D-21, a 12.8m long drone, made from the same materials as the A-12 and equipped with a ramjet engine. The D-21 would be launched at 80,000ft and Mach 3 and would be even faster than the A-12.

In December 1964, the D-21 was mounted on an A-12 (now called the M-12) and made its first flight.

After take-off, the mission required the M-12 to be refueled in flight by a KC-135. After an in-flight collision, Kelly Johnson canceled this project in 1966.

The A-12 gave rise to the YF-12A, a model designed for high-speed interception missions. Only in 1965, when it was publicly announced, was the A-12 recognized as the owner of nine world speed records including a speed of 2,000 mph and flight at an altitude of 80,000 ft.

A total of 18 aircraft were produced, 13 A-12s, 2 M-12s and 3 YF-12As. One of the A-12s had a rear seat for pilot training and continued to operate with J-75 engines throughout its life.

Three more A-12s were subsequently lost. On July 9, 1964, an aircraft crashed during the final approach for landing when a roll control servo froze at an altitude of 500 feet and 200 knots causing the aircraft to make a smooth, steady movement to the left. Lockheed test pilot Bill Park was unable to prevent it. With a pitch angle of around 45 degrees and an altitude of 200 feet, he ejected and was thrown sideways out of the aircraft. Despite the low altitude, his parachute opened in time, and he landed safely.

On December 28, 1965, the third A-12 was lost when it crashed 30 seconds after takeoff when a series of rapid and violent yaws made the aircraft uncontrollable. Mele Vojvodich was scheduled to make a performance check flight that included in-flight refueling in a KC-135. He managed to eject safely 200 feet above the ground.

A post-accident investigation revealed that the main cause of the accident was a maintenance error. An electrician had mistakenly changed the electrical connections between the Stability Augmentation System's gyroscopes and the servos on the control surfaces. As a result, the longitudinal and yaw commands were inverted.

The investigation pointed to the electrician's fault but also pointed to a failure to supervise maintenance and the fact that the design of the aircraft allowed the connection to be changed in the first place.

The SR-71 "Blackbird" was a longer and more fuel-efficient version of the A-12. It entered service in January 1966 and remained in service for 33 years. A total of 32 SR-71s were built, but 12 were lost in accidents.

Its primary mission was reconnaissance. But when an anti-aircraft artillery threat was detected, the procedure was to accelerate the aircraft and escape the missile. Missiles could not climb to the altitude where the Blackbirds were flying with enough fuel to accelerate and hit them.

The windshield of the SR-71 was made of quartz and fused to the titanium structure using ultrasonic fusion. During flights, and even during landing, the windshield reached temperatures of over 300° C.

The SR-71's landing speeds were over 170 knots, which required the use of parachutes to help the aircraft decelerate.

The SR-71's fuselage panels were fitted with plenty of ground clearance to allow for thermal expansion of several centimeters. But this and the lack of fuel sealants to withstand the thermal expansion and extreme temperatures meant that fuel leaked onto the ground before take-off, much to the annoyance of the mechanics.

The Blackbird's engines were designed for optimum performance at Mach 3.2. This was achieved through the air intake. It had a cone called a spike. At subsonic speeds, the spike was all the way forward. But as the aircraft accelerated supersonically, the position of the cone retracted so that the shock wave formed was tangential to the air inlet, preventing air from leaking out of the inlet. When the cone was adjusted correctly, the shock wave was captured inside the air intake and the air intake was said to be "Started." In addition, another shockwave occurred inside the engine to decelerate the air to subsonic speeds in the diffuser, so that the air entered the compressor at the correct speed.

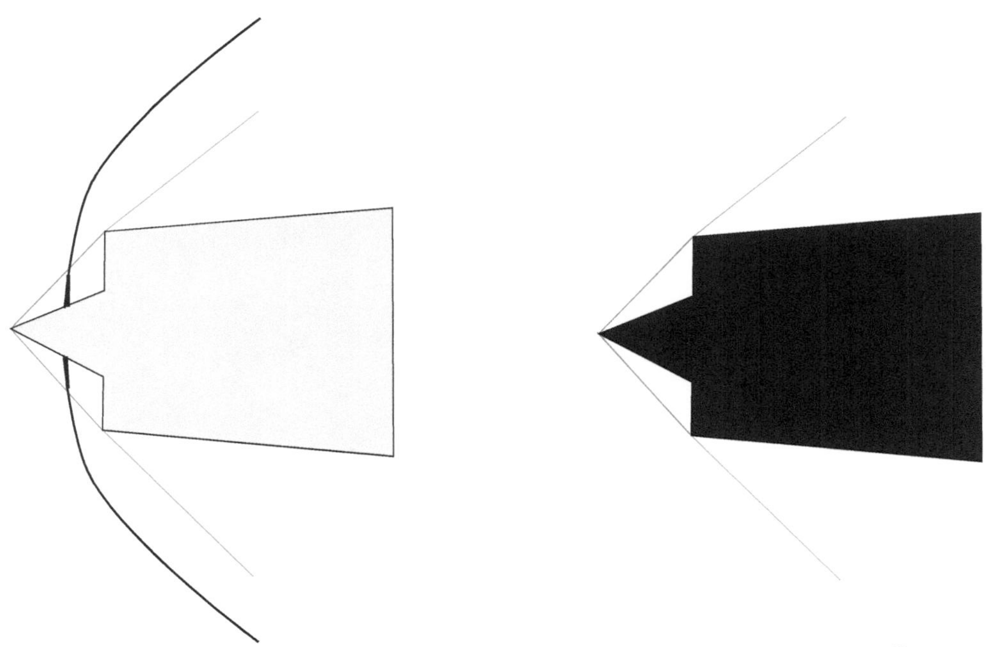

Air intake not started (left) and started (right) at Mach 2.0[19]

[19] Internet Archive NASA Glenn Research Center. Public Domain.

On take-off, only around 25% of the engines' thrust came from the afterburners, but at Mach 3, almost 100% of the thrust came from the afterburners. Part of the air from the compressors was removed and directed directly to the afterburners via bypasses. The engines flew continuously in afterburner, except during in-flight refueling.

Triethylburane (TEB) was used to start the SR-71's engines, which ignited on contact with air, producing a green flame. The planes were refueled after take-off because the aircraft's maximum speeds were only possible with in-flight refueling and to avoid excess weight on take-off. Specialized KC-135Q tankers were needed to refuel the SR-71. The KC-135Q had a modified high-speed boom, which allowed the Blackbird to be refueled at almost the maximum speed of the refueler.

One problem with the propulsion system was that the analog computers sometimes could not adjust in time to the changing conditions, and one of the engines lost its air intake setting (i.e. "became unstarted"). This usually caused the afterburner to cut out, followed by a strong yaw moment due to the asymmetrical traction of the engines. The autopilot, stability augmentation system (SAS), and the pilot would try to correct this yaw, but it could also cause the other engine to stall, followed by loud noise and vibration.

After wind tunnel tests and computer models, Lockheed implemented an electronic control system to prevent this phenomenon and restart the air intake without pilot intervention.

During this investigation, it was also discovered that vortices originating from the "chines" of the fuselage were sometimes entering the engines. The "chines" were the aerodynamic elongations of the aircraft's fuselage, made of composite material.

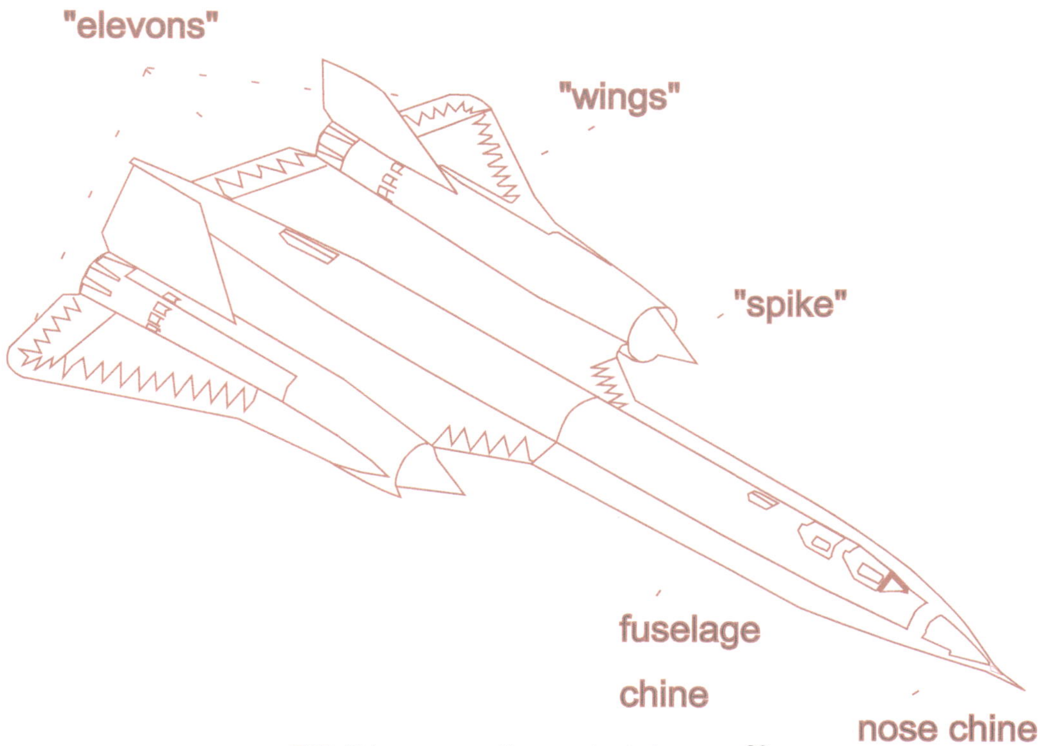

SR-71 composite material areas[20]

Flying at 80,000 feet meant that crews could not use standard masks, as they could not provide enough oxygen above 43,000 feet.

Pressurized protective suits were produced for the crew by the David Clark Company. In addition, an emergency ejection at Mach 3.2 would subject the crews to temperatures of around 230°C. In the event of a high-altitude ejection, an oxygen supply would keep the suit pressurized during descent.

The cabin could be pressurized at an altitude of 10,000 or 26,000 feet during flight but needed a high-demand cooling system since flying at Mach 3.2 heated the outside surface of the aircraft to around 260°C and the inside of the windshield to around 120°C.

The air-conditioning system used a heat exchanger to cool the cabin and heat the fuel at the same time. The same air-conditioning system was also used to keep the forward landing gear compartment cool, so eliminating the need for special aluminum-impregnated tires similar to those used on the main landing gear.

[20] United States Air Force Pilot Handbook. Public domain.

The pilots of the Blackbird and the reconnaissance officers at the rear post were given food and drink for the reconnaissance flights, which were often long. The water bottles had straws which the crew members guided into an opening in their helmets by looking in a mirror. Similarly, food was kept in containers like toothpaste tubes, which brought the food to the crew member's mouth through the opening in the helmet.

The Blackbirds usually made one flight a week due to the time required to prepare the missions. This incredible aircraft was in service with NASA until 1999. No Blackbird was ever shot down.

FLIGHT TEST PROCESSES

3

"To hit the center of the target is not the most important thing, but the process to launch the arrow."
Ancient Japanese tradition

Flight Test processes involve several areas of the organization and are, therefore, important and highly visible. Most of them must be controlled by Test Engineering because these engineers are responsible for defining the tests that will be carried out, and reporting problems identified in the aircraft.

FLIGHT TEST PROCESSES

These are repetitive activities that take place within the Flight Test Organization. The processes affect all the organization's deliverables directly, or indirectly.

The processes must guarantee agile, safe, and repeatable work focused on:
- Operations safety
- The organization's goals

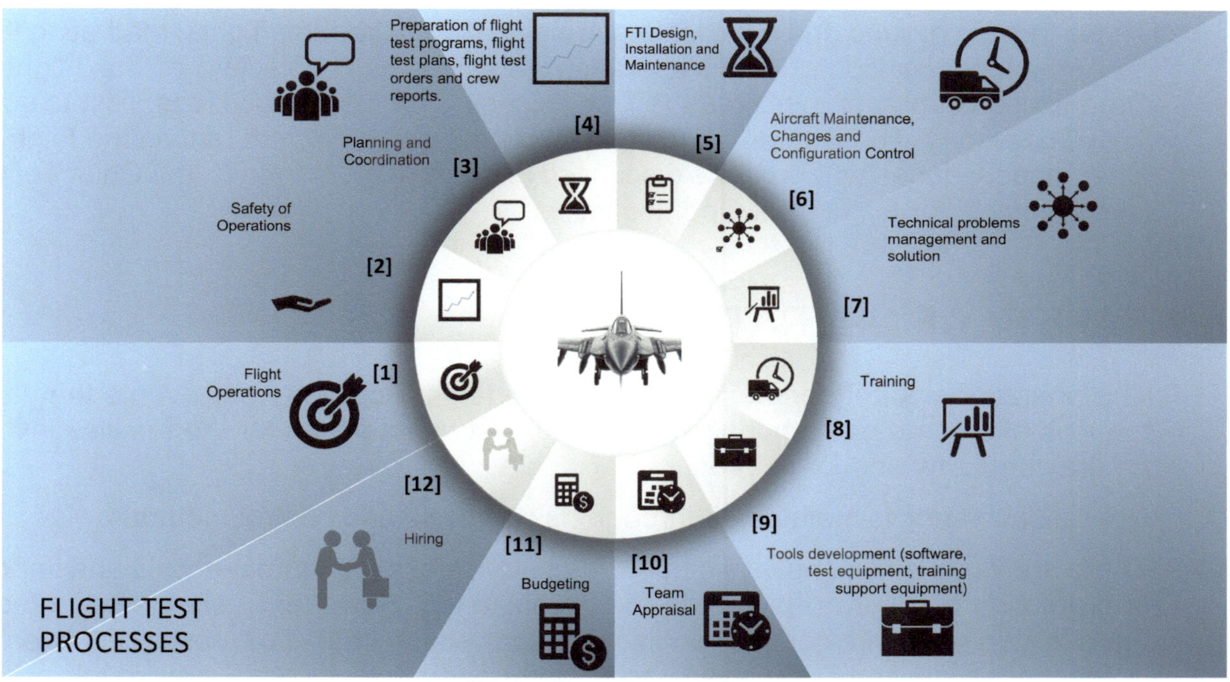

Flight Test Processes

The main processes are:

1. **Flight Operations**
2. **Safety of Operations**
3. **Planning and Coordination of Activities**
4. **Preparation of test plans, test orders, test reports, and data analysis**
5. **FTI definition, design, installation, and maintenance**
6. **Aircraft maintenance, changes and configuration control**
7. **Reporting and controlling aircraft technical problems**

8. **Training**
9. **Tools Development (computers, test equipment, training support, etc.)**
10. **Team Appraisal**
11. **Budgeting**
12. **Hiring**

All the twelve processes occur at the same time There are interactions among these 12 processes that affect results. For example, a difficult process for preparing Test Orders could lead to delays in other processes. This is why these processes must be simple and efficient so as not to overload the team, causing problems such as reduced efficiency, relationship problems, health or psychological problems, and so on. Doctrine teaches that the deadlines are always very "at the limit".

FLIGHT OPERATIONS

These include ground and flight testing, training flights, transfer flights, marketing flights, and maintenance flights. Flight operations are the main objective of the Flight Test Organizations and are therefore the first process. The other processes exist to support Test Operations. These operations must follow the guidelines set forth in the Flight Test Operations Manual (which will be discussed in Chapter 4). Normally, the Chief Test Pilot is responsible for Flight Test Operations.

SAFETY OF OPERATIONS

The operational safety process prevents accidents. Accidents are events that involve the loss of life and equipment, with enormous consequences for the families involved, for the company, and for its products.

Operations safety is maintained through risk analysis and control methods.

There are various risk analysis models available, and all of them are valid and important in promoting a culture of safety in the organization. However, it should be noted that there is little point in a comprehensive risk analysis process if the result of this process is not applied to Flight Test Orders, and latterly, to each test flight. To avoid this problem, there needs to be a filter that directs a general and comprehensive view of the risks (all the existing risks in the operation) to a concise and operational view of a single flight.

Chapter 12 will discuss the Safety of Operations during Flight Tests in details.

PLANNING AND COORDINATION

For the Flight Test Organization to fulfill its mission within the company, it needs to manage several projects. These projects must be managed using the Flight Test Organization's processes. The main tools for this are:

a. Short-term planning
b. Long-term planning
c. Program Meetings
d. Clearance to Start or Resume Testing

Short-term meeting

Short-term meetings plan activities in a short space of time of around two weeks. This allows face-to-face alignment with those involved in each activity, and each test, and ensures the necessary support. The activities planned in Short-Term Meetings must be accessible, at any time, to all the team members taking part in the meeting, and others who are involved in the activities. The best way to do this is with an application that allows everyone to view aircraft calendars, that is easy to use and flexible such as Microsoft Outlook.

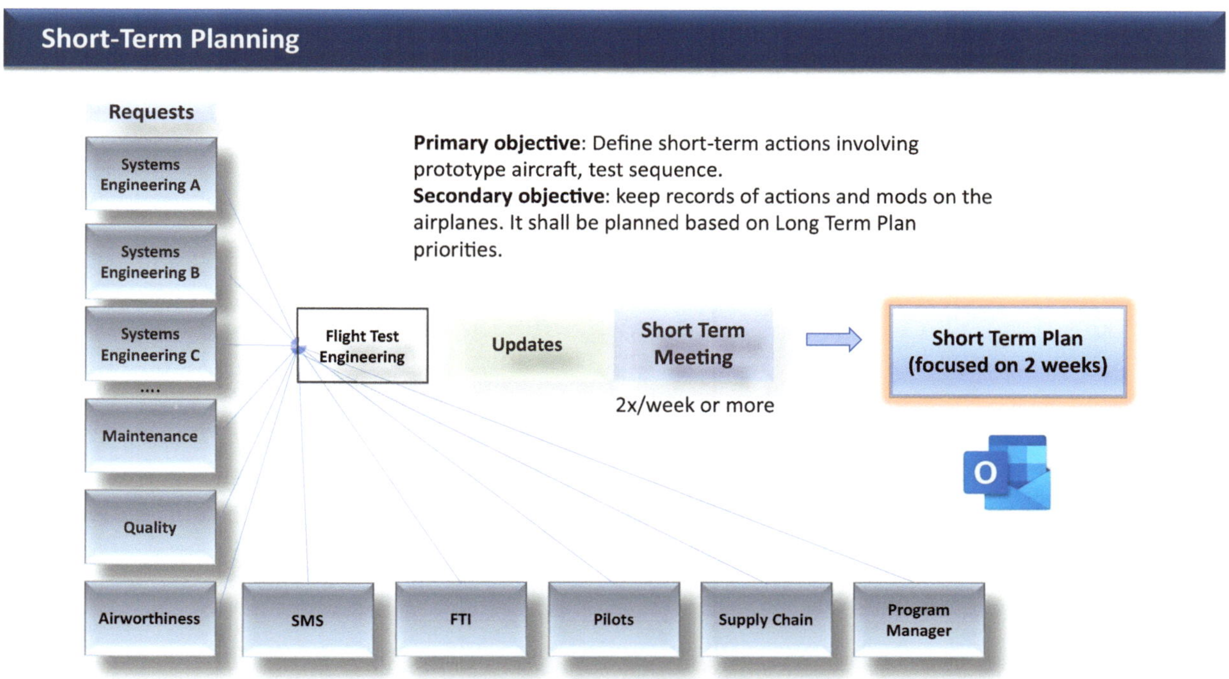

Short-term planning

Long-term meeting

Long-Term Meetings provide a long view that in principle covers the entire length of the project, anticipating the actions and resources needed so that Short-Term Tests can be carried out. These meetings must be held for each engineering team separately. The recommended tool for all long-term planning is Microsoft Project.

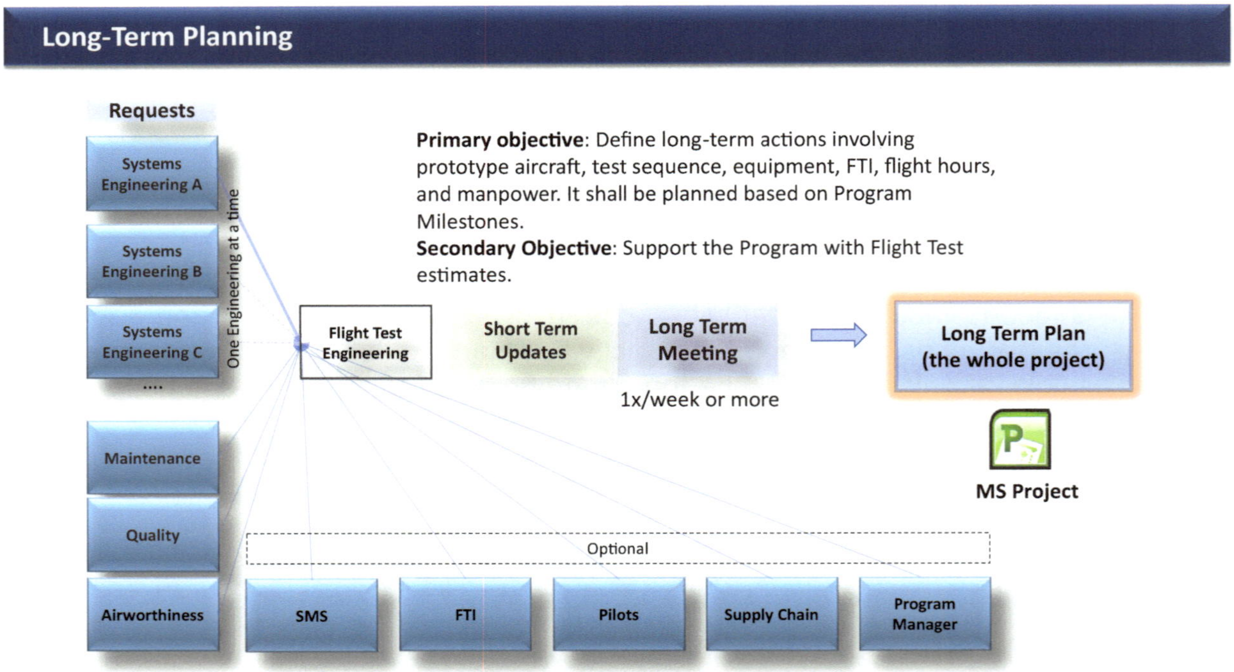

Long-term planning

Program Meetings

Program meetings are where the major areas of the organization are aligned. Flight Testers must attend these important meetings and report any difficulties and the progress of the test campaign.

Clearance to Start or Resume Testing

Clearance to start a campaign requires that the Test Plan has been approved, among other preparations which normally include:

- Aircraft preparations for flight
- Aircraft modifications
- New Flight Test Instrumentation
- Documentation Approval
- Safe of Flight justifications
- Meetings
- Permit to Fly from Aviation Authorities.

There is often also a meeting to discuss updates and ensure that the necessary actions have been taken. This meeting is called Test Readiness Review.

Sometimes a sequence of tests is often stopped to find solutions to problems. Once corrective measures have been adopted, it is necessary to return to activities. This is when "checkpoints" are made. Checkpoints (also called Test Readiness Reviews) confirm that the necessary prerequisites for resuming a given test have been met. Examples include:

- Aircraft and test documentation
- Changes to the aircraft
- Instrumentation
- Compliance

This meeting acts as a kind of checklist so that the tests can proceed. The advantage of these checkpoints is the guarantee that the test can proceed safely and that everyone involved is aware of the existing conditions and limitations.

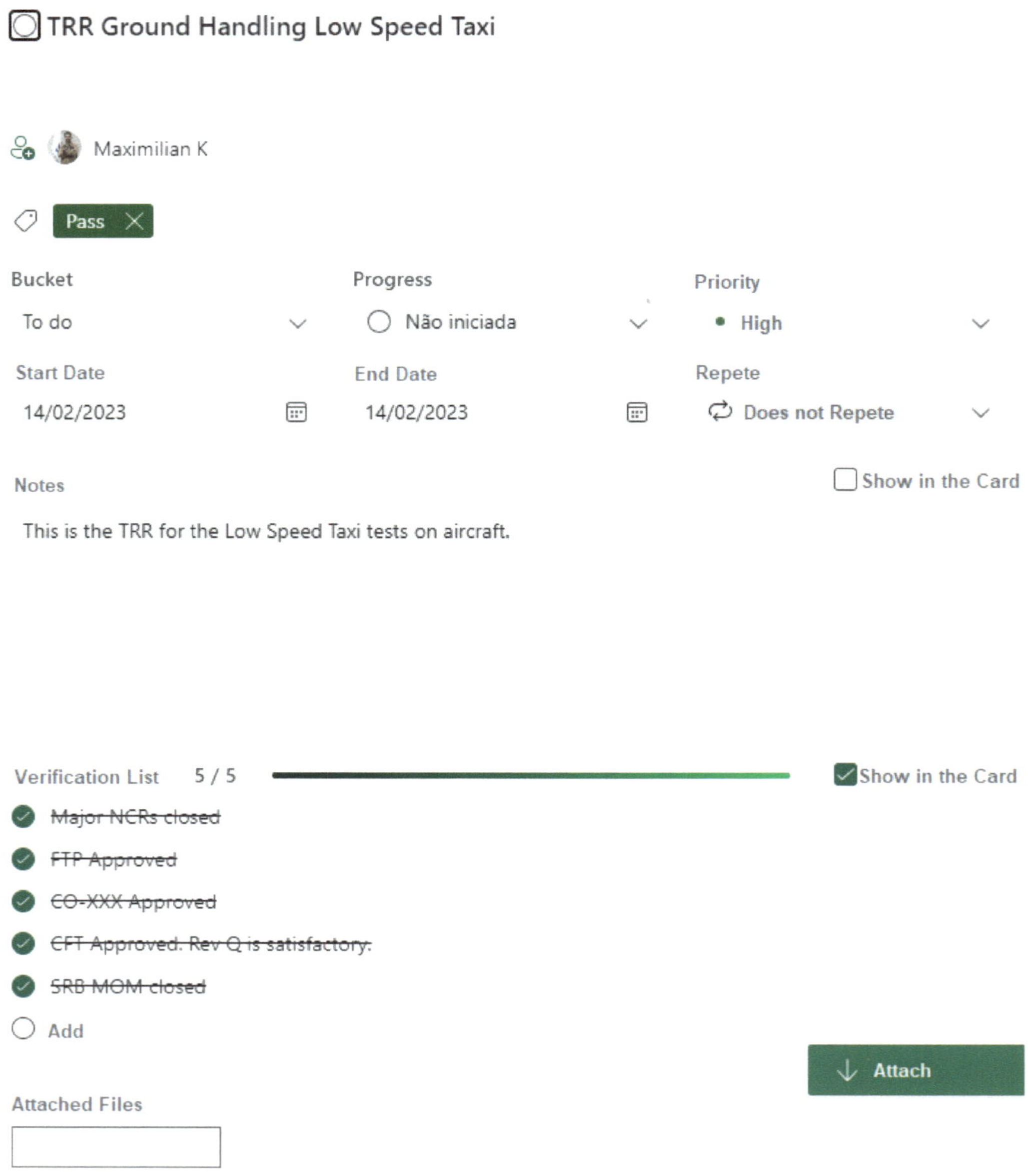

Example of "Trial Readiness Review" (done in MS Planner)

Clearance for First Flight

The release for the first flight of a new aircraft must be treated as a special case. It requires additional preparation and coordination:

- Definition of the aircraft configuration
- Identification of potential aircraft problems and threats
- System rigs, tests carried out, and results
- Simulation of the aircraft's flight characteristics
- Preparation of support logistics to the first flight
- Crew training
- Instrumentation review
- Safety Equipment Review
- Resources to support the first flight (telemetry, fire department, ambulance, ATC, other airspace and aerodrome users, etc.)
- Tests that will be carried out on the aircraft before the First Flight, and a review of their results
- Human Factors evaluation
- Definition of the Phases for Release for the First Flight
- Phase Change Meetings for the First Flight
- Preparation of the First Flight Test Plan
- Preparation of the First Flight Risk Analysis
- Permit to Fly (PtF) approval

PREPARATION OF TEST PLANS, FLIGHT TEST ORDERS, FLIGHT TEST REPORTS, AND DATA ANALYSIS

Preparation of Flight Test Program Plans

Flight Test Programs are documents that contain broad terms describing what will be tested by one or more aircraft. They are very useful for planning the material, logistical, and personnel efforts required for long campaigns. They are also requested by aviation authorities and may be necessary for the issuance of a Permit to Fly (PtF).

Normally, a plan is prepared that contains the types of tests to be performed for each aircraft without going into too much detail about execution dates, as these can change significantly.

It is advisable to allow a margin in the estimated resources, such as flight hours and number of professionals, for example. As some of these estimates are not relevant to the authorities, they can be removed from this report.

The Flight Test Program is a deliverable of the Flight Test Organization.

Preparation of Flight Test Plans

Flight Test Plans detail what needs to be tested by the Flight Test Organization. They are, therefore, the basic units that make up the total set of tests to be carried out.

Flight Test Plans must contain at least:

- A Test Matrix (set of test points to be carried out). It must be very detailed to avoid missing test points, or those carried out under the wrong or incomplete conditions.
- Maneuvers to be executed, with their description.
- Required aircraft configurations (weight and CG, aerodynamic configuration, etc.).
- Required FTI. Defining the required FTI can be a major effort with long lead time and require modifications to the aircraft, the development of new calibration methods, the calibration of sensors outside the company, etc.
- Test risk analysis. Must be carried out as early as possible so that there is time to respond and take action. The method recommended in FAA Order 4040.26, which uses the definition and mitigation of *hazards* associated with testing, is one of the methods that will be covered in Chapter 12, Risk Assessment Methods. The Three Flags method, develop by the author, will also be covered in chapter 12. The best place to put the Risk Analysis is in the Flight Test Plan.

Also, the preparation of the Test Plans must follow specific steps which are:

- A draft Proposal of the test made by System Engineering
- A Resources Review meeting to discuss the resources required for the tests (FTI, cameras, tail chute, special authorization, etc.)
- A Technical Review meeting to discuss the procedures and limitation of the tests.
- A Safety Review Meeting to discuss the hazards associated with the tests.

The Test Plan should not be excessively complicated or long to facilitate its preparation and avoid mistakes.

Preparation of Test Reports

The result of each test carried out on the prototype aircraft must be consolidated in the form of a Flight Test Crew Report.
However, the format of this report can be very different among Flight Test Organizations.

When trying to use a report format from another type of organization (using the format of a Teaching Organization in industry, for example) many problems will occur.

Problems arise because the format taught and advocated at Flight Test Schools is very detailed, making it difficult to use in industry. It is quite common for organizations to try to use the "Seven Part Paragraph" method. Trying to follow this method, learned in training schools, has the opposite, undesirable effect for the following reasons:
- The team does not have the necessary resources to prepare such a long report and keep up the pace of flights.
- The engineering team also lacks the resources needed to analyze a lengthy report, process the data recorded during the flights, and maintain support for subsequent test flights.

Note: Special Flight Test Report
Situations, where problems that affect the safety of operations are discovered, a more detailed report containing opinions and data may be necessary so that the problem can be formally recognized by the rest of the company.

This is the function of the Special Flight Test Report. Unfortunately, few Flight Test Organizations have this important mechanism. Problems encountered during the test campaign are reported, but the lack of a dedicated report can mean that the problem takes longer to be acknowledged and fixed.

Preparing Flight Test Orders

Flight Test Orders are documents that define in full detail how a particular test is to be carried out. They must contain:
- Test objective
- Trial limitations

- Aircraft Configuration
- Aircraft Instrumentation
- Testing areas
- Test frequencies
- Aircraft weight and center of gravity
- Test Points (with a description of procedure, altitude, speed, weight, center of gravity, flaps, landing gear, systems)

Flight Test Orders must be uniquely and sequentially numbered for each aircraft. This makes it easier to find a relevant fact or piece of information. Preferably, this number should contain the prototype number, the type of document (in this case, Test Order), and a test sequence number.
Flight Test Orders are made up of "Flight Cards." Each flight card is a detailed procedure applicable to a given flight condition and maneuver. The order of the Flight Cards is also very important and must be in the intended sequence of execution. This prevents errors and makes it easier to keep track of what has been carried out.

The numbering of ground tests and flight tests must be separated. This is to facilitate control of the number of tests performed under these conditions. Unique numbering will cause enormous difficulties in generating activities reports.

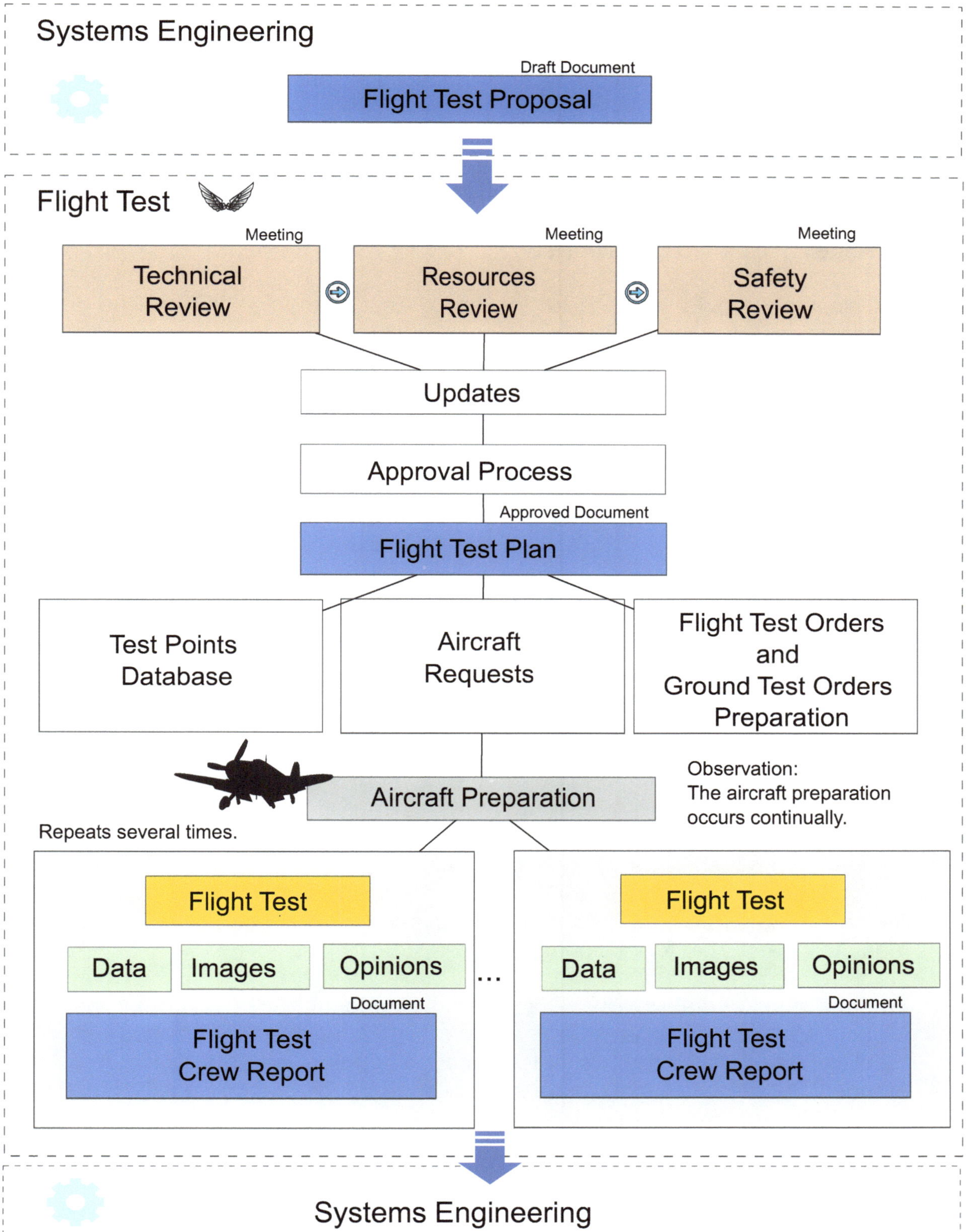

Preparation of Flight Test Plans, Flight Test Orders, and Flight Test Crew Reports

DEFINITION, DESIGN, INSTALLATION, AND MAINTENANCE OF FLIGHT TEST INSTRUMENTATION (FTI)

Flight Test Instrumentation is one of the most important tools to be prepared before the start of a test campaign. The definition of the FTI is, therefore, an important part of the tests, and requires a great deal of effort from various areas (mainly FTI, Engineering, and Maintenance).
It is not only necessary to define the FTI but also to design, purchase, and define when and how the FTI will be installed and calibrated during the Test Campaign. The FTI work package is, therefore, substantial.

FTI planning should be done for the entire Test Campaign. FTI will be discussed in Chapter 8.

AIRCRAFT MAINTENANCE

Aircraft maintenance involves dozens of activities and requires tremendous effort and dedication. It is a regulated activity that must follow the principles of supervision, repeatability, and experience. Maintenance planning must be done for the entire Test Campaign. Aircraft maintenance will be dealt with in Chapter 9.

REPORTING AND MANAGING TECHNICAL PROBLEMS

The main objective of Flight Tests is to ensure that new aircraft developed meet their design and certification requirements and serve their customers. However, this will not be achieved in every case. Problems will often arise, forcing design modifications to be made and retested. This is why a system for reporting and controlling these problems is necessary.
Ideally, there should only be one source of problem reports, but in general technical problems are reported through more than one source. Can be cited:
- Pilot reports to maintenance.
- Reports from test crews via flight test crew reports.

As these problems accumulate over time, it is necessary to use a control tool. There are many tools available for this on the market.

The tool should be able to store problems by system or by ATA[21] 100 system and also be able for users to enter comments and suggestions, attach images, etc. One very practical way of keeping track of technical problems is to use a tool such as Microsoft Planner. But there are many others, such as the JIRA System. It is not recommended to use MS Excel for this purpose.

---

[21] Air Transport Association.

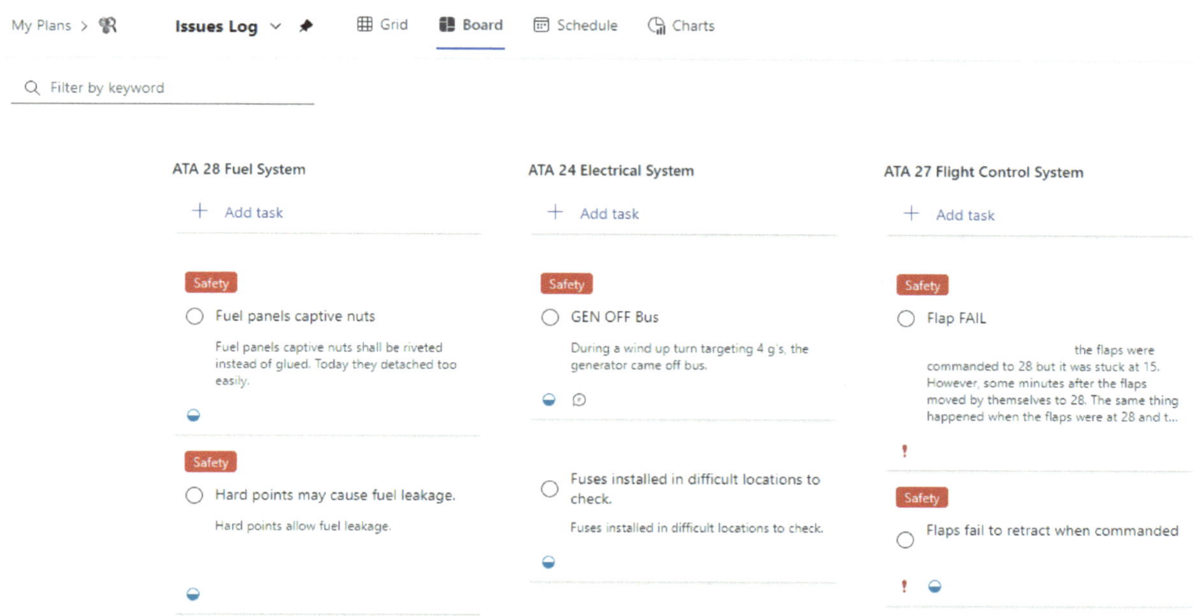

Technical problem management (done in Microsoft Planner)

TRAINING

This chapter deals with the training required for Flight Test crew members.

There are three main types of training:

- Process Training
- Training in Tools (computer tools, equipment, etc.)
- Operational Training
  - Testing Technique Training (in classroom or on the aircraft)
  - Ground Training (in classroom or on the aircraft)
  - In-flight training
  - Engineering Simulator Training
  - Flight Simulators training

All training must be evaluated using a Training Evaluation Form.

The trainee must fill in and sign the Training Record Sheets that are kept by the company.

Training must be carried out initially (when a new crew member or employee is hired) and repeatedly to ensure operational performance.

The training should be defined in an internal Flight Test Area standard so that there are no doubts, and so that it is repetitive.

The importance of training

Training helps improve efficiency and repeatability and improves safety and the quality of results. In addition, training is the best way to ensure that company procedures are followed correctly by all employees.

The importance of supervision

Supervision is the best way to guarantee the effectiveness of training. The Training Supervisor should be a Lead FTE, or their delegate who will oversee the quality of the training, making sure that the right standards are achieved, and procedures are followed.

It is often not possible to find the number of professionals needed to carry out a Development Testing Campaign in the market. The solution to this is usually training.

Training can be internal or external through courses. Internal training can end up increasing the workload of the more experienced project teams, but the return on investment soon becomes evident.

Training plays a major role in the safety of testing operations. But how can good training be guaranteed? We can mention (see figure below):

- Preparation time.
- Existence of instructors. It is the instructors who will guarantee good training. To have good instructors, it is important to have pilots and engineers with instructional experience on the team. However, this is not usually a priority when setting up a Flight Test team.
- Customization. Training needs to be adapted to the characteristics of the company and its products so that each crew member is prepared to carry out their duties.
- Resources for Instruction. Aircraft, classrooms, teaching materials, engineering simulators, rigs, mockups, etc.

Resources needed for good training

Due to the lack of training resources, it is common for organizations to send their crew members to take training courses at recognized schools. Despite the high costs, this is one of the most common solutions to this problem and it usually gets good results. External training involves a lot of anticipation and resources but has the advantage of bringing new knowledge to the team.

However, they have the disadvantage of not being fully applicable to the prototype aircraft that will be used in the Test Campaigns. It is this training, customized for a specific aircraft and test campaign, that really makes the difference and translates into higher flight safety.

TYPES OF TRAINING

Training in Flight Test processes

Flight Tests work with many different processes. Each new test campaign needs to be planned, executed, and reported on. The processes aim to ensure the maturity and safety of the Test Campaigns while meeting the needs of internal and external customers.

Topics to be covered:

- Flight Test Processes
    - Planning and carrying out test flights
    - Short-term meetings

- Long-term meetings
- Technical Review Boards
- Meetings to review testing systems and techniques
- Safety Review Meetings
- Telemetry procedures

- Flight Test Pilot Processes

    - Initial training
    - Recurrent Training (IFR, Simulators, CRM, hypobaric chamber, sea survival, etc.)
    - Operational Safety Training (HSE, Accident Prevention)
    - SMS

- Flight Test Operation Processes

    - Operational Procedures Training.
    - Escape Systems Training (ejection seats, ejection procedures, lifeboats)
    - Operating emergencies
    - Crew Resource Management (CRM)

Training in Tools

The Flight Test Organization uses various software tools to carry out flight tests. Knowledge of these tools is essential for carrying out tests safely and avoiding errors.
Examples:
- Real-time data visualization tools (e.g. IADS, IMC Famos, etc.)
- Flight Test Portal
- Post-flight data analysis tools (e.g. MATLAB, Scilab, Excel)
- Weight and CG calculation tools
- Trial Planning Tools (Microsoft Project)

Operational Training

To be able to carry out flight tests, it is essential to have extensive knowledge of the test aircraft and of the safety systems, limitations of the aircraft, and operating procedures followed by the company. This training can be divided into two parts: Ground Training and Flight Training.

Ground Training

- Airport emergency procedures
- Telemetry procedures

Preferably described in a company norm. The roles of the telemetry members and how they interact with each other and with the test aircraft must be clearly defined.
- Ejection seat training
- Operational procedures at the test airport
- Test areas
- Aircraft Operational Limitations
- Aircraft cockpit training
- Airplane Flight Manual (AFM)
- Flight Test Instrumentation
- Flight Test Techniques
- Aircraft Systems (or Aircraft Ground School)

In-flight training

The company's training standard should clearly state how many and which flights are required for initial and recurrent Flight Test crew training. It must be specific and state the content of each flight clearly and in detail.

Another important aspect is training in flight test techniques. Test techniques must be taught and trained initially and recurrently to ensure correct execution during real test campaigns.

Training in Engineering Simulators

Training on engineering simulators is a very practical and inexpensive tool for training engineers and pilots on a new aircraft. In addition, in the early stages of aircraft development, it is an important tool for evaluating the cockpit layout and the aircraft's normal and emergency procedures.

There are different types of engineering simulators, depending on the stage of development of the aircraft:
- Cabin mockups
- Conceptual simulators
- Systems integrators
- Flight quality simulators

Flight Simulator Training

Flight simulator training is a very practical and safe tool for training engineers and pilots on an already certified aircraft. It also helps to reduce training costs. Simulator training is recommended in average once a year for all crew members.

Training Evaluation

Trainees must be formally evaluated during each training session. The purpose of the assessment is to consolidate knowledge and evaluate the results of the training.

The training must be evaluated by the trainee using a Training Evaluation Form. Training Evaluation Forms must be kept by the organization.

Training records

Training must be recorded using Training Record and kept in the organization's Training Records.

Required Skills for Flight Test Engineers

"Soft Skills (A)	1. Communication
	2. Teamwork
	3. Time Management
	4. Adaptability to Change and Different Cultures
	5. Follow Procedures Strictly
	6. Self-learning
	7. Leadership
Technical Skills (B)	1. Aerodynamics and Flight Mechanics
	2. Aircraft Systems
	3. Flight Test Techniques (theory)
	4. Written Communication
	5. Project Management
	6. Programming
	7. Flight Test Risk Management and HSE
	8. Data Analysis
	9. FTI systems
	10. Aviation Requirements (FAR 23/25, MIL-STD-8785, MIL-HDBK-244, MIL-HDBK-1763)
Operational Skills (C)	1. Cockpit training
	2. Piloting
	3. Flight Test Techniques

Required Skills for Flight Test Pilots

"Soft Skills (A)	1. Communication
	2. Teamwork
	3. Time Management
	4. Adaptability to Change and Different Cultures
	5. Follow Procedures
	6. Self-learning
	7. Leadership
Technical Skills (B)	1. Aerodynamics and Flight Mechanics
	2. Aircraft Systems
	3. Flight Test Techniques (theory)
	4. Written communication
	5. Flight Test Risk Management and HSE
	6. Aviation Requirements (FAR 23/25, MIL-STD-8785, MIL-HDBK-244, MIL-HDBK-1763)
Operational Skills (C)	1. Cockpit training
	2. Piloting different types of aircraft
	3. Flight Test Techniques
	4. Operations in the Cold/Desert/Over Water/Combat
	5. Instrument flight
	6. Technical English

AIRCRAFT MODIFICATION AND CONFIGURATION CONTROL

Precise configuration control of each aircraft is necessary. Preferably, it should be done by an online tool. This control must be led by the Configuration Control Team, executed by the Maintenance and FTI Team, and checked by the Quality and Project Team. The Configuration Control Team should be responsible for authorizing changes to the project and all associated documentation. The Maintenance Team is responsible for implementing these changes and recording them. The Quality Team must double-check between what has been installed or removed and what has been recorded.

Configuration control must also be carried out using continuous monitoring of all modifications made to the aircraft through meetings and a database. Thus, every time a new modification is made to the aircraft, it must be entered into this database.

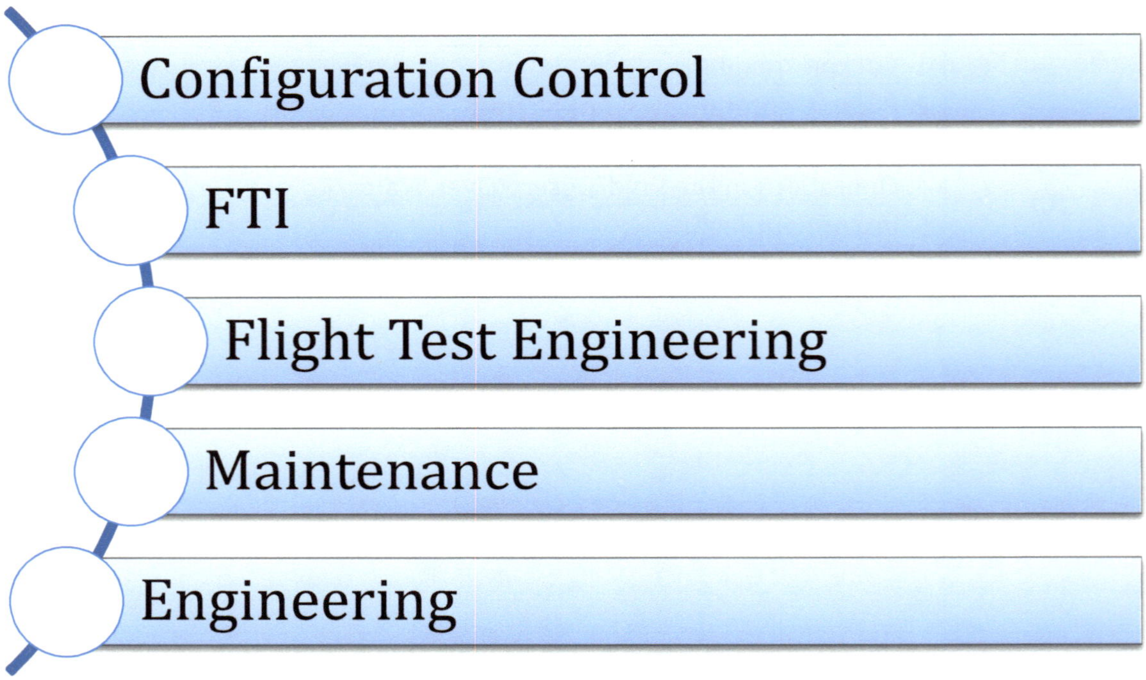

Prototype Configuration Control Team

TOOLS DEVELOPMENT

The analysis of test data, although a primary function of Systems Engineering, should also be carried out by the Flight Test team. This requires the development of data reduction tools, which are often commonplace, and facilitates a healthy exchange of information between the areas involved, without one area being too dominant." The exchange and cross-referencing of data enables a more in-depth analysis, benefiting problem-solving. There are several data analysis software tools on the market that can be used for this.

Data analysis tools can be used for real-time or offline analysis.
The most used real-time analysis tools are:

- Mathworks MATLAB/Simulink
- Curtiss Wright / IADS
- IMC / Famos

The main offline analysis tools are

- Mathworks / MATLAB
- Curtiss Wright/IADS
- Microsoft Excel
- Dassault Systems / Scilab
- OriginLab / Origin

Alongside data analysis tools, there are data visualization software. They can be real-time or offiline.

Some companies prefer to develop their own data visualization and analysis software, but these are exceptions.

Offline Analysis Software (post-flight)

Of the software above, only Scilab is freeware. This is a great advantage, as it not only results in cost savings, but also makes it easier for the whole team, interns, and teams from other areas to use the tools. So, the effort can be shared, and more robust tools can be developed.

On the other hand, powerful tools such as MATLAB, Origin, and Famos can reduce analysis time and produce results that would otherwise take a long time. Therefore, the decision of which *software* to use depends on the application and the resources available in the company.

In the field of flight testing, heavy computational tools are not usually necessary. Those that can give a quick response and produce preliminary test results to be used in subsequent flights, thus improving analysis and increasing safety, are required.

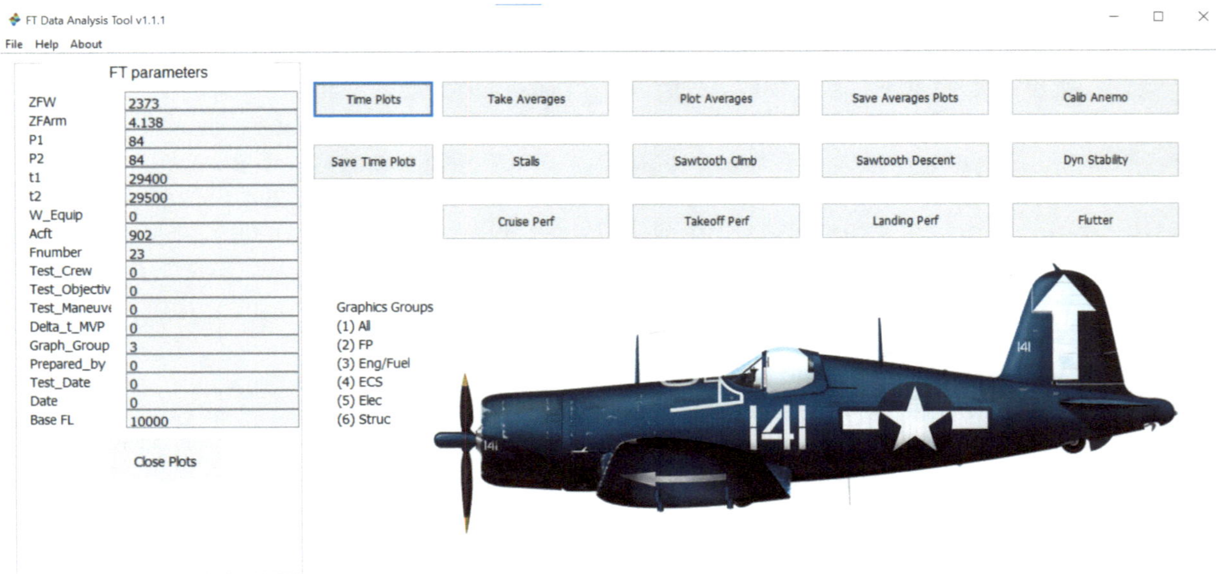

Example of post-flight data analysis software (done in Scilab)

Real-Time Analysis (in-flight)

Among the software mentioned above, Curtiss Wright/IADS, MATLAB/Simulink, Curtiss Wright/IADS and IMC / Famos can be used in real time. Real-time programming is very useful for monitoring limits and calculating results. This type of program is extremely important in the development of new aircraft where practically all design limits will be tested.

TEAM APPRAISAL

Team appraisal is necessary for all teams, to improve performance, identify points for improvement, and, above all, identify high-performance points to be copied.

Evaluating people's performance must be based on facts and data. For example, during the preparation of a Stoll Testing Campaign, engineer So-and-So was responsible for a Technical Review Board meeting. The way the meeting was conducted, and its results are good material for an assessment. The assessment should not be based on impressions or assumptions. That is why day-to-day events are relevant material for evaluations.

Regular activities should also be evaluated. For example, a regular meeting can be evaluated for its effectiveness, duration, function, results obtained, and so on.

Regular reports should also be evaluated. For example, a weekly report can be evaluated for its content (poor or excessive) and relevance (is it worth doing, or not?).

BUDGET MANAGEMENT

Budget management must be done regularly and in advance so that there is time to request additional resources and to identify additional resource needs as early as possible. The budget needs to have an approval cycle to be validated. Furthermore, the participation of people involved in the planning of the Test Campaigns is fundamental. Based on this, an estimate of the number of working hours should be made to support hiring decisions. This estimate should consider:

- The number of aircraft to be tested
- The number of flight hours to be flown
- The duration of each program
- The number of people needed

By adding up all these activities and estimating the number of working hours required for each activity, it is possible to calculate the total working hours required. This can also be done using software, such as Microsoft Project. It is not recommended to do this in a Microsoft Excel spreadsheet because of the high chance of error. In addition, Microsoft Excel requires many more hours of work to prepare and update correctly. Do not work for a software, make a software work for you.

HIRING

Hiring must be an agile process. To this end, the skills required for each job offer must be well defined and each given a weight. This must be done by the hiring department.

Interviews with candidates should be carried out after a set of questions has been defined so that there is no lack of information after the interview, and everything runs smoothly. This should be done in conjunction with the company's Human Resources department.

Hiring qualified staff will also depend on the company's ability to offer an attractive salary and benefits. A technical interview and a managerial interview are usually carried out.

The purpose of the technical interview is to ascertain the candidate's technical experience for the position. The technical interview is an opportunity to give a first impression of the company professionalism and projects to the candidate.
The managerial interview is designed to give the candidate who passed the technical interview the opportunity to meet the managers or directors of the area where they will be working.

After the two interviews, the final phase of salary and benefits discussions begins, usually led by Human Resources.

NOTE ON CRISES MANAGEMENT

A crisis can be defined as an event that generates demands on the company. The size of the crise is proportional to the size of the demands. There are two types of crises: a crises generated by Organizational Milestones, and a crise generated by Safety Events. Crises happen, be prepared.

Crises generated by Project Milestones

Project Milestones can generate enormous pressure on the company because they are associated with contracts and, sometimes, directly with major clients. That is why it is important to have a mechanism associated with these crises that can provide quick responses and solutions. For example:
- Write a report or document in response to the crisis generated
- Gather data to substantiate an aircraft capability being questioned
- Make flights with clients to demonstrate the aircraft's capabilities

Crises generated by safety events

If an event occurs that affects the safety of an aircraft or its crew, another type of crisis is generated. This crisis must be managed by a trained team with support from senior management. It is also necessary to use a Crisis Management Manual with the necessary instructions for such circumstances. These crises are beyond the control of the Flight Test Organization. These are very difficult situations that generate a lot of pressure, delays, and additional costs. Be ready for this.

FLIGHT TEST OPERATIONS MANUAL

4

"The secret of achievement is to hold a Picture of a successful outcome in your mind."
Henry Thoureau

The *Flight Test Operations Manual* (FTOM) is an essential part of the *Flight Test* Organization and defines the structure, processes, and responsibilities within the organization.

The EASA requirement governing the existence of a Flight Test Operations Manual is 21.A.143, which appears in Subpart J of CS 21 for Aeronautical Design Organizations (EASA) or the equivalent 21.A.243 for Production Organizations when they have Flight Test operations.

There is, therefore, a correspondence between the Flight Test Operations Manual and the Design Organization Exposition. Both are related and must be consistent with each other.

These requirements demand that the Flight Test Operations Manual specifically contain:

1. A description of the organization's Flight Test processes, including its involvement in the Permit to Fly issuing process.

2. Crew policy, including composition, competence, recurrence, and flight time limitations.

3. Procedures for the transportation of persons other than crew members and for Flight Test training, where applicable.

4. Security risk management policy, and associated methodologies.

5. Procedures for identifying equipment and instruments to be carried on board.

6. A list of documents that need to be prepared to carry out Flight Tests.

The Flight Test Operations Manual must be adjusted for each organization according to its size, purpose, and type of operations. It can even form part of the Project Organization's Manual.

The EASA Manual for Flight Test Operations Manuals is a guide to the development of this manual. It was first issued in April 2018, and its purpose is to guide Aircraft Design Organizations (see Chapter 15) and Aircraft Production Organizations (see Chapter 16) to develop their manuals. It is only a guide and not a regulation on the subject.

CHAPTERS OF THE FLIGHT TEST OPERATIONS MANUAL

FLIGHT TEST PROCESSES

If the FTOM is presented as a separate document, it must include a chart indicating the structure of the organization and, more specifically, the functional links of those responsible for flight test activities.

It should also mention coordination between all the departments that affect Flight Testing, e.g. the Design Office, Production and Maintenance, in particular coordination for the establishment and updating of a Flight Test Program.

CREW POLICY

According to the Flight Test category, the FTOM should describe the organization's policy on crew composition and the competence and status of its flight test crew members, including procedures for appointing crew members for each specific flight.

All crew members must be listed in the FTOM.

A flight time limitation policy should be established.

TRAINING

The FTOM must describe how Flight Test training is done.

The qualification of the flight test crew can be guaranteed through recent experience or refresher training.

For aircraft where CS 21 Appendix XII applies to (the ones which have a Permit to Fly), the minimum flight experience per year shall be:

- For pilots: 50 hours.

In addition:

- For pilots with a flight test qualification, the 50 hours must include 20 hours of flight tests in any Flight Test category.

- For pilots conducting category 3 flight tests, the minimum experience must be expressed in terms of the number of flights conducted prior to the issue of a Certificate of Airworthiness (CA).
- For pilots carrying out category 4 flight tests, the minimum experience must be proportional to the activity envisaged.

- For Lead FTE (LFTE): 10 hours of flight tests in any Flight Test category.

The FTOM must specify the requirements for refresher training to ensure that crew members are sufficiently up to date to carry out the required Flight Test activities.

A system must be established to record the effectiveness of Flight Test crew training.

A valid national document (i.e. a license), issued by an EASA Member State supported by national regulations and guaranteeing compliance with the agreed upgrade requirements, is an acceptable means of compliance to demonstrate the upgrade of a pilot holding a Flight Test qualification and for a Flight Test engineer.

PERSONS ON BOARD WHO ARE NOT QUALIFIED CREW MEMBERS

According to the flight test category, the FTOM must describe the organization's policy regarding the presence and safety on board of persons who are not crew members (i.e. without on-board duties).

Persons other than crew members must not be allowed on board Category 1 flight tests.

RISK MANAGEMENT

The FTOM should describe the organization's policy on risk and safety assessment, risk mitigation, and associated methodologies. Chapter 12 deals with Risk Management in Flight Testing in details.

EQUIPMENT IDENTIFICATION PROCEDURES

The FTOM must list, depending on the nature of the flight, the specific safety-related instruments and equipment that must be available in the aircraft or carried by those on board.

The FTOM must contain provisions allowing flights to be carried out in the event of defective or missing instruments or equipment.

LIST OF DOCUMENTS

The FTOM must list the documents to be produced for the Flight Trials, and include (or reference) the procedures for issuing, updating, and following them up to ensure control of the documents:

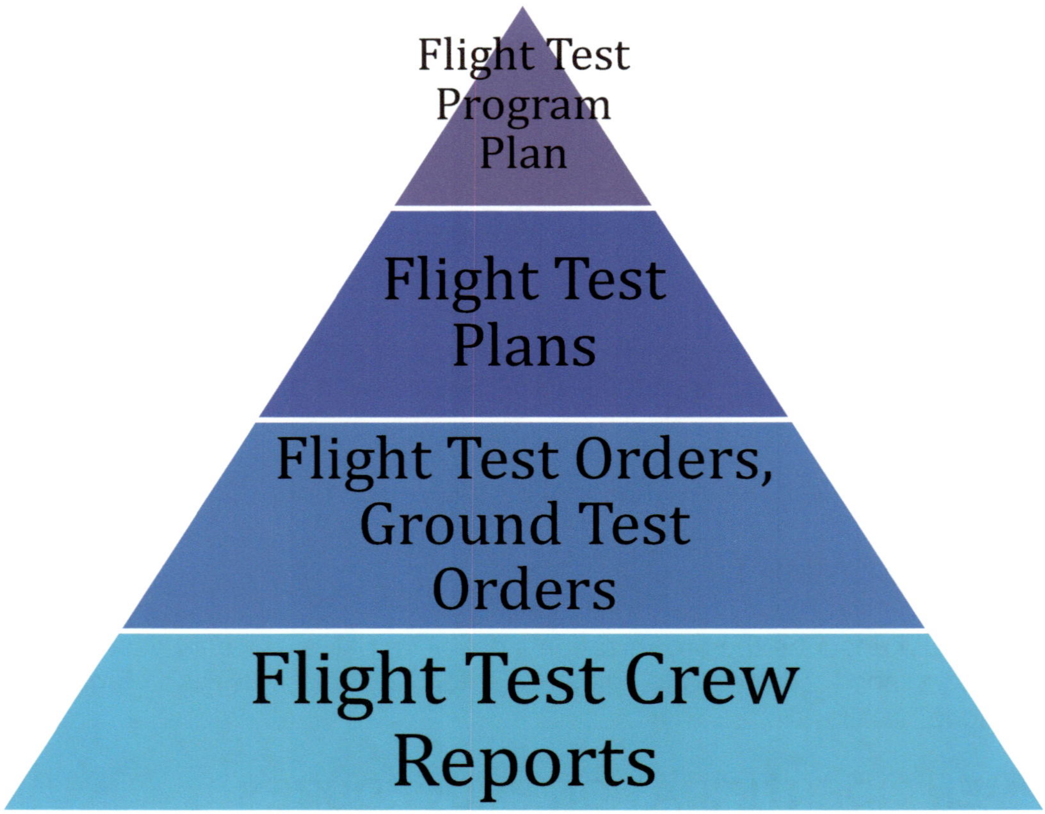

Document Structure of a Flight Test Organization

(i) documents associated with a Flight Test Program:

- Flight Test Order, for a given flight, which must include:

    - A list of the tests to be carried out and the associated conditions
    - Safety considerations relevant to the flight
    - Flight category (e.g. category 1)
    - Crew composition
    - Names of people who are not crew members
    - Aircraft configuration items relevant to the test will be highlighted to the crew
    - Loading the aircraft
    - Reference to approved flight conditions
    - Limitations relevant to the flight to be highlighted to the crew

- Flight test crew report.

(ii) documentation and information to be carried on the aircraft during the flight test

(iii) record keeping: the FTOM must describe the policy on record keeping.

FLIGHT PERMITS

The FTOM must describe the involvement of the Flight Test Organization or Flight Test team (as appropriate) in the process of approving flight conditions and issuing Permits to Fly per FAR/CS 21 Subpart P.

FLIGHT TESTS CARRIED OUT BY SUBCONTRACTORS

When flight tests are carried out by contractors or subcontractors, they must comply with the FTOM of the primary organizations, unless they have established an FTOM per Part 21, the use of which has been agreed between the two organizations.

FLIGHT TEST ENGINEERING

5

"People who say it cannot be done should not interrupt those who are doing it."
George Bernard Shaw

Flight Test Engineering is the area responsible for planning and conducting Flight Test Campaigns. This entails a big responsibility, often not understood by the rest of the company. Communicating correctly with the rest of the company is the only way to manage this problem.
To be able to plan properly, Test Engineering requires a strong theoretical and practical basis from its engineers.
To be able to do this, Test Engineering requires considerable operational experience.
This is why Test Engineering is where the operational world meets aircraft design and production.

RESPONSIBILITIES

Flight Test Engineering is the area normally responsible for:

- Planning the Flight Test Campaigns to achieve the company's milestones.
- Coordinating resources for the execution of Test Campaigns.
- Discussing the preparation of the campaigns with the pilots, Engineering, Certification, Program, Maintenance, and Instrumentation.
- Carrying out test flights as a crew member on board the aircraft or conducting the tests in coordination with the pilots via telemetry.
- Preparing and using tools and techniques to guarantee the validity of test data, if possible, in real time.
- Prioritizing safety when conducting Flight Test Campaigns.
- Following the company's processes set out in the Flight Test Operations Manual.

The main deliverables of Flight Test Engineering are:

1. Preparing and executing the Flight Test Program. The Flight Test Program is a high-level document that presents the demands for tests that need to be carried out. When this demand is well defined, Test Engineering can write a Flight Test Program containing:
    - Program Objectives.
    - Number of prototypes required.
    - Estimated total flight hours per year.
    - Estimated number of flight hours per prototype per year.
    - Special equipment required (e.g. flutter vans, electrical load banks, ballasts).
    - Off-site testing campaigns.
    - Training flight hours.
    - Transfer flight hours.
    - Chase aircraft flight hours.
    - Marketing/Demonstration of flight hours.
    - Number of FTE, pilots, FTI, and Maintenance Team members.

An important detail is that it is not necessary to contain a timetable with dates, as this quickly becomes obsolete, but rather a logical sequence of Test Campaigns and their estimated duration. Dates are not necessary to estimate the resources required.

2. Leading Resources Reviews. These meetings are part of the Flight Test Process and aim to determine sufficiently in advance all the resources that will be required for a given test. The results of this meeting must be recorded and each assigned to a responsible person. When a campaign requires a great deal of resource coordination, several Resource Review meetings may be necessary (e.g. Weapons Campaigns). The FTE leading the campaign is responsible for coordinating these meetings and making sure that the resources are being provided.

3. Leading Technical Reviews relating to Flight Test Campaigns. The Technical Review Meetings are part of the Flight Test Process and aim to analyze the tests to be carried out, procedures, and expected results.

4. Leading Safety Reviews relating to Flight Test Campaigns. Safety Review Meetings are part of the Flight Test Process and aim to identify threats to test safety.

5. Preparing Flight Test Plans. Using the information obtained at the Flight Test Process meetings, an FTE is responsible, normally along with a Systems Engineer, for writing and approving a Flight Test Plan. This document should be the only reference for carrying out the tests to ensure consensus among everyone.

6. Preparing Flight Test Orders. Flight Test Orders are prepared by the FTE, reviewed and approved by another FTE, or by pilots.

7. Taking part in test flights as crew. Flight Test Engineers have a duty to take part in test flights as aircraft crew members.

8. Writing Flight Test Crew Reports. One of the duties of Flight Test Engineers is to write Flight Test Reports.

9. Writing Flight Test Campaign Reports. It is often not done, but writing Flight Test Campaign Reports is of great value to the company. This type of report usually contains many lessons learned and the most relevant results from a long period of testing. It is done by Flight Test Engineers.

10. Preparing and executing Short-Term Planning. Short-term planning brings together the main customers in a Test Campaign to make decisions about the campaign, activities required, delays, resources, dates, and deadlines. A Flight Test Engineer is usually responsible for this. The frequency of these meetings can be a few times a week, or daily.

11. Preparing and carrying out long-term planning. Good long-term planning is not an easy task. It requires good knowledge of software tools, such as Microsoft Project. It also requires detailed knowledge of all the tests that will be carried out and their expected duration. However, without this planning there is no reasonable forecast for

all the activities and when they will begin and end. Essential. Made by Flight Test Engineers.

12. Attending meetings about technical problems with the aircraft. The purpose of Flight Trials is to identify possible problems with the aircraft. When these problems exist, they need to be dealt with as quickly as possible, as the cost of solving them increases with time. These meetings are conducted by other areas of the company, but they require information on the problems raised, which is provided by Flight Test Engineers, or Flight Test pilots.

13. Attending prototype Change and Configuration Control Meetings. Prototype configuration control affects flight safety and must, therefore, be closely followed by Test Engineers. They are usually held weekly.

14. Preparing Weekly Flight Test Reports. These inform the rest of the company about the tests carried out, and their main results. They are not formal documents, but short presentations, preferably with lots of pictures (which are worth a thousand words). Your main client is the Program.

15. Participating in training. Training should be recurrent to maintain the proficiency of the Flight Test teams.

16. Preparing Special Flight Test Reports when necessary to explain important events that occurred during the tests in more detail. They may be unnecessary if Systems Engineering can make an analysis and produce a similar (formal or informal) report in good time.

17. Preparing software for data analysis, parameter calculation, and alarms in telemetry.

18. Leading telemetry teams during test flights.

19. Participating in recurrent training in flight test techniques.

20. Developing data analysis tools during and after flights.

To be able to make all these deliveries, the Test Engineering team needs to be very well trained and organized and have a great deal of technical knowledge.

FLIGHT TEST ENGINEERING TEAM

The Flight Test Engineering team normally consists of:
- One chief engineer or manager who is administratively responsible for the entire team, interfacing with the Head of Flight Tests, Human Resources, and other areas of the company, such as Engineering and Programs.
- Lead engineers ("Lead FTE") who are responsible for leading the activities of the company's various programs.
- Flight Test Engineers, who will be responsible for conducting flights and Flight Test Campaigns.

Division of the Flight Test Engineering Team by Programs

The deliverables mentioned above are specific to each program and are, therefore, repeated for all of them, which greatly increases the team's workload.

TEAM SPECIALIZATION

To make it easier to divide up the numerous Flight Test Campaigns and allow the team as much preparation time as possible, it is very useful to divide up the Test Engineering team technically. This also has the advantage of being able to provide specific training according to the campaigns that will be carried out by the FTEs. This increases the quality of the results and the safety of the Test Campaigns.
A practical way of dividing the team is shown in the figure below.

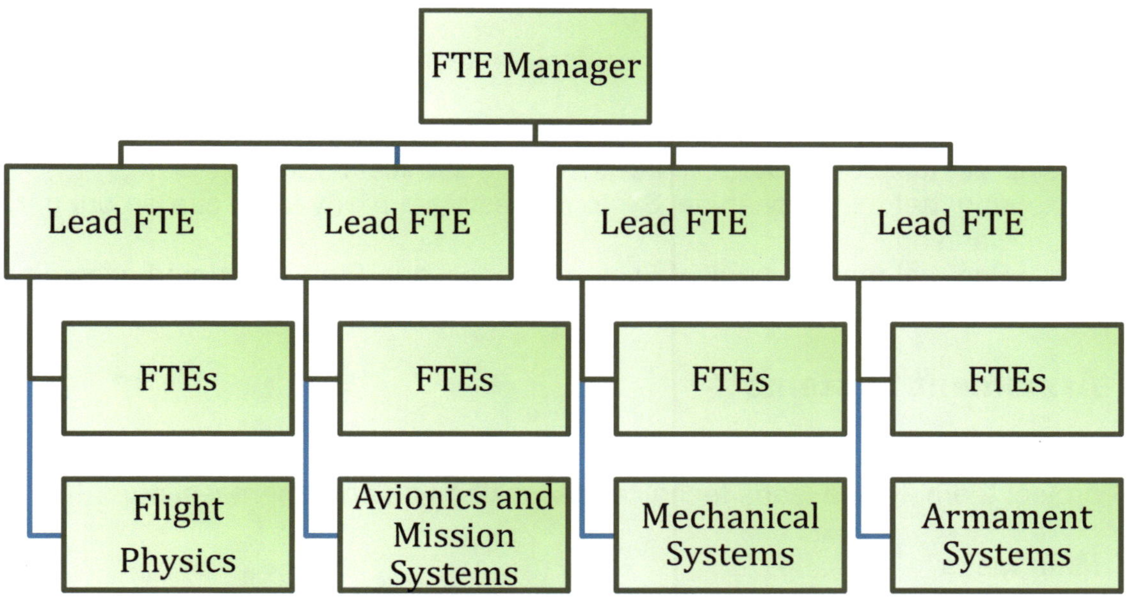

Division of the Flight Test Engineering Team by Specialty

Flight physics

Plan, prepare, conduct, and report Flight Tests related to the Flight Physics engineering project team. Also lead technical discussions on flight physics tests.

Main tasks
- Being the test coordination link with the Flight Physics team.
- Ensuring that the Flight Physics tests are ready to be carried out per the Short-Term Planning.
- Making sure that the Flight Physics Long-Term Plan is updated according to the latest information.

Avionics and Mission Systems

Plan, prepare, conduct, and report on Flight Trials related to the Avionics and Mission Systems engineering project team.
It also leads technical discussions on testing avionics and mission systems.

Main tasks

- Being the test coordination link with the Avionics and Mission Systems team.
- Ensuring that the Avionics and Mission Systems are ready to be deployed according to plan.
- Ensuring that the Avionics and Mission Systems Long-Term Plan is updated in line with the latest information.

Mechanical Systems

Plan, prepare, conduct, and report on Flight Tests related to the Mechanical Systems project team.
In addition, it leads technical discussions on testing mechanical systems.

Main tasks

- Being the test coordination link with the Mechanical Systems team.
- Ensuring that the Mechanical Systems Tests are ready to be carried out per Short-Term Planning.
- Ensuring that the Mechanical Systems Long-Term Plan is updated according to the latest information.

Armament Systems

Plan, prepare, conduct, and report on Flight Trials related to the Armaments Engineering project team. It also leads technical discussions on armament testing.

Main tasks

- Being the test coordination link with the Armament Systems team.
- Ensuring that the Armament Tests are ready to be carried out according to the Planning.
- Ensuring that the Long-Term Armament Systems Plan is updated according to the latest information.

FLIGHT TEST PILOTS

6

"What makes the boat move is not the billowing sail, but the wind that cannot be seen."
Plato

Flight test pilots are the heart of Flight Test Organizations. They have the difficult task of flying the prototype aircraft in all types of tests, often putting their lives at risk. This is why these risks must be very well assessed, as described in Chapter 12.

Normally, pilots are needed who have a great deal of operational experience combined with a strong theoretical background. A good example is Commander James Doolittle, whose life story was discussed in Chapter 2 of this book. These are the objectives of the Flight Test Schools covered in Chapter 1. However, Flight Test pilots are responsible for several other activities, which we will describe into more details below.

RESPONSIBILITIES

- Carrying out test flights as commander or co-pilot of a prototype aircraft.
- Providing support for the preparation of the Aircraft Flight Manual.
- Providing support for Human Factors Design. These activities are wide-ranging and involve defining the aircraft's avionics system, panels, alarms, power handle, visibility, ejection seat, etc.
- Supporting negotiations between the company and its clients when necessary. This applies to company meetings with clients, reviewing operational requirements, etc.
- Participating in technical discussions with Certification Authorities. These discussions usually involve compliance with requirements in which the operational view is essential.

FLIGHT TEST CATEGORIES

The Flight Test categories have been defined by EASA in its "*Easy Access Rules for Airworthiness and Environmental Certification Regulation*" EU Number 748/2012 Appendix XII. There are four categories of Flight Tests:

CATEGORY 1

(a) Initial flights of a new type of aircraft or an aircraft that has been significantly modified.
(b) Flights during which the potential encounter of flight qualities significantly different from those already known is anticipated
(c) Flights to investigate new or unusual design features
(d) Flights to determine or expand the flight envelope
(e) Flights to determine performance, flight characteristics and piloting qualities when the limits of the flight envelope are reached
(f) Training flights for category 1 flights.

CATEGORY 2

a) Flights not classified in category 1 on aircraft not yet certified
(b) Flights not classified in category 1 on an aircraft already certified, after the incorporation of a modification not yet approved and which:

(i) require an assessment of the general behavior of the aircraft or

(ii) require an evaluation of basic crew procedures when a new or modified system is operating or is required or

(iii) are required to intentionally fly outside the limitations of the currently approved operational envelope but within the investigated flight envelope.

(c) Training flights for category 2 flights.

CATEGORY 3:

Flights carried out to issue a declaration of conformity for a newly-built aircraft that do not require flying outside the limitations of the aircraft's type certificate or flight manual (typical case of production flights).

CATEGORY 4:

Flights not classified in category 1 or 2 in an aircraft of an already certified type in the case of implementing a design change that has not yet been approved. This is the case for design changes that do not alter the flight characteristics or flight envelope of the aircraft. Good examples are "*Minor*" modifications to Type Certification.

Test flights on aircraft that have not yet been certified must be classified as category 1 or 2.

FLIGHT TEST QUALIFICATION AND FLIGHT TEST PILOT COMPETENCY LEVELS

There is no consensus among the world's aviation authorities as to whether it is mandatory to regulate the profession of Flight Test Pilot. In 2013, the FAA canceled FAA Order 8110.41A, which dealt with the subject. However, EASA published its "*Easy Access Rules for Flight Crew Licensing FCL*" in August 2020. This regulation contains requirements for Flight Test pilots and Flight Test instructors.

Holders of an aircraft or helicopter pilot's license may only act as pilot-in-command in flight test category 1 or 2, as defined in Part 21 when they hold a flight test rating.

The issue of a Flight Test Rating requires:

(1) holding at least a CPL (commercial pilot) license and an IR (instrument rating) in the appropriate aircraft category.

(2) having completed at least 1,000 flight hours in the appropriate aircraft category, of which at least 400 hours as PIC.

(3) have completed a training course in an Approved Training Organization (ATO) appropriate to the intended aircraft and flight category. The training must cover at least the following subjects:

- Performance
- Stability and control/Handling qualities
- Systems
- Flight Test Management
- Risk/safety management

The privileges of holders of a Flight Test rating may be extended to other flight test categories and other aircraft categories upon completion of an additional training course at an ATO (*Approved Training Organization*).

According to EASA Part FCL, Flight Test pilots may have the following qualifications:

Level 1:
Perform all categories of Flight Tests, as defined in Part 21, as pilot-in-command or co-pilot.

Level 2:
Perform category 1 flight tests as defined in part 21:
- as co-pilot, or
- as pilot-in-command, in the case of CS23 aircraft or aircraft with equivalent airworthiness and weighing no less than 2,000kg, except those in the *commuter* category, or with a maximum dive speed above Mach 0.6, or a maximum ceiling above 25,000 feet.

Perform all other categories of Flight Tests, as defined in Part 21, as pilot-in-command or co-pilot

The requirements for level 3 and 4 flight test pilots are defined by the UK Civil Aviation Authority (CAA) as follows:

Level 3:
Hold a valid license appropriate to the category of aircraft being tested, issued per Part FCL, and hold a Commercial Pilot's License (CPL) as a minimum. In addition, the pilot-in-command must:
(a) hold a Flight Test rating, or
(b) have at least 1,000 hours of flight experience as pilot-in-command in aircraft of similar complexity and characteristics, and
(c) have participated, for each class or type of aircraft, in all the flights that form part of the program leading to the issue of the individual certificate of airworthiness for at least five aircraft.

Note: The certificate of airworthiness is issued after the production flights of an aircraft. The typical example of level 3 test pilots are pilots who perform production flights.

Level 4:
Hold a valid license appropriate to the category of the test aircraft, issued per Part FCL, and hold at least a CPL. The pilot-in-command must hold a test flight qualification or have at least 1,000 hours as pilot-in-command in aircraft of similar complexity and characteristics.

ORGANIZATION OF THE TEST PILOTS

Pilots are usually led by a Chief Test Pilot, who reports to the Head of Flight Test (usually a director-level position in the company).
The chief test pilot usually needs to divide his team according to the company's programs, designating a leader for each program, and the pilots who will service that program.

In addition, the chief test pilot assigns a team of pilots (or a single pilot) to participate in the development and evaluation of the cockpit philosophy, ergonomics, field of vision, etc. This is known as the Human Factors project.

And finally, the Chief Test Pilot, with the Head of Flight Tests, designates a pilot as the Air Safety Officer. This pilot has the important role of ensuring flight safety and operations, advising the Head of Flight Tests and the entire team.

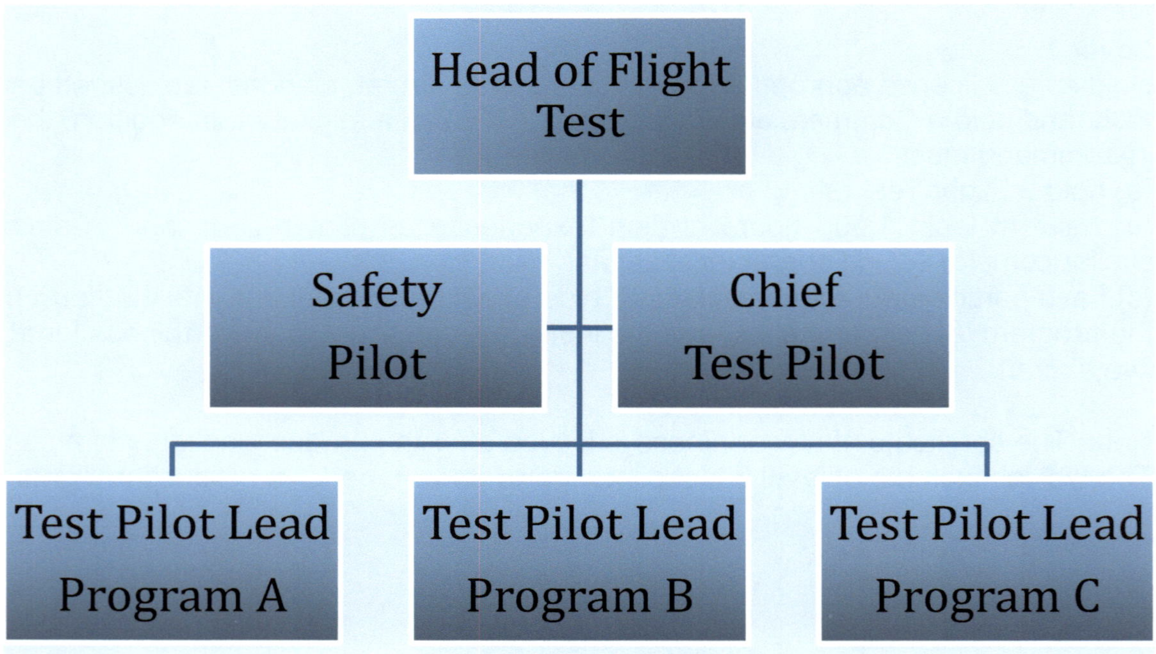

Organization of Test Pilots

Within each program, it is up to the lead pilot to choose which pilots will carry out the particular Flight Test Campaigns. This choice is very important for increasing efficiency, data quality, and safety because each pilot, regardless of their EASA qualification level, usually finds it easier to carry out certain types of tests. Personal characteristics such as experience, patience, knowledge of avionics systems, piloting precision in performance maneuvers, flying qualities, or weapons maneuvers are key factors in this decision.

MILITARY AND CIVILIAN PILOTS

It should be noted that there is a significant difference between the training of military and civilian pilots, and this is true all over the world. Military pilots have experience in the following types of flying that civilian pilots do not normally have:
- Formation flights
- Attack and defense missions
- Weapons missions
- Mission equipment (datalinks, FLIR, self-defense, etc.)

Civilian pilots, on the other hand, have experience in the following areas that military pilots do not normally have:
- Cabin Resource Management (CRM)
- Experience in dealing with passengers
- Experience in dealing with customers
- Experience with large aircraft
- Experience in long-haul flights

These differences and others translate into attitudes, disciplines, working methods, and leadership. Sometimes a military pilot is better suited to a particular test campaign and sometimes a civilian pilot is better suited. Sometimes it takes a military pilot and a civilian pilot to bridge these differences.

Of course, personal characteristics are also often decisive.

FLIGHT OPERATIONS SUPPORT

7

"The most difficult thing in life is decide which bridge to cross and which to burn."
Bertrand Russel

A Flight Test Organization is complex even when small and can involve several flights a day, flights in several different countries, transatlantic flights, demonstrations in several countries, and numerous teams.
To help deal with this complexity, an Operations Support Sector is highly recommended (see Chapter 1, Typical Organization Chart).

This has several positive consequences:
- It allows the technical team to concentrate on discussions and preparations for the tests from a technical point of view
- It allows the technical team to have more time for other tasks such as coordination with the Engineering, Maintenance, and Program teams.

RESPONSIBILITIES

- Coordinating test flights with organizations outside the company (ATC, operators at the same aerodrome).
- Coordinating with support groups for tests (firefighters, rescue vehicles, ambulances, helicopters, etc.).
- Coordinating the necessary support from Fixed Based Operators (FBOs).
- Coordinating participation in shows and demonstrations with the Event Organizers.
- Checking the validity of crew licenses and training.
- Coordinating the obtaining of overflight authorizations and customs clearances.
- Disseminating operational procedures that affect air operations.

Crew training is also often part of Operations Support duties, which makes this area even more relevant and necessary.

QUALIFICATIONS

There are no strict requirements for professionals working in Operations Support. However, experience in aviation is a must. Whether as a pilot, flight controller, or supporting the operations of other companies.

FLIGHT TEST INSTRUMENTATION

8

"The result of all the battles is decided before the battle begins."
Sun Tzu

Flight Test Instrumentation consists of equipment for collecting data from an aircraft. It plays a fundamental role in aircraft development and certification, without which it is not possible to identify and correct aircraft problems and demonstrate certification requirements.

To perform this function, Flight Test Instrumentation (FTI) needs to be designed long before the first flight of a prototype aircraft. In general, this takes one to two years but can vary depending on the number and type of parameters required. The starting point of the FTI project is the definition of all the necessary parameters.

DEFINITION OF FTI

Defining the FTI is a long process that involves all of the company's systems engineering departments. Due to its great complexity and number of clients, it is very difficult to have all the FTI installed, calibrated, and tested before the Test Campaign begins. For this reason, a good practice is to define a base FTI that meets a basic set of tests and, from there, expand the list of valid parameters through activities carried out on the aircraft little by little. This defines the so-called "deltas" that will be needed for well-defined Test Campaigns. For example, you know that a fuel flow meter will be needed for the Cruise Performance Campaign. This sensor can then be defined as an "FTI delta," and sufficient time must be planned for its definition, purchase, calibration, and installation so that it is operational before the Cruise Performance tests.

The approach described here, therefore, aims to divide the total work into smaller packages (called Delta FTI) to be installed and calibrated on the test aircraft. In addition, define a list of baseline FTIs that must be operational and calibrated for the duration of the test campaign. For each subsequent test campaign, a delta FTI must be defined and installed (along with the baseline FTI list).

So, the FTI must be defined in 5 steps:

1. Definition of FTI Maps by Engineering Customers. FTI review meetings must be held to determine the list of FTI parameters for each engineering system. Each subset of the FTI List is called an FTI Map. The complete list of FTI parameters is called the FTIL (Flight Test Instrumentation List).
2. Definition of a Baseline Map (minimum to support the initial flight test campaigns).
3. Definition of the Delta FTI packages to be installed before each test campaign. Delta FTIs are smaller packages to be installed so that the test campaign is not greatly affected by modifications to the FTI.
4. Initial definition of FTIL.
5. FTIL approval.
6. Final FTIL.

When a new FTI is requested, the cycle repeats.

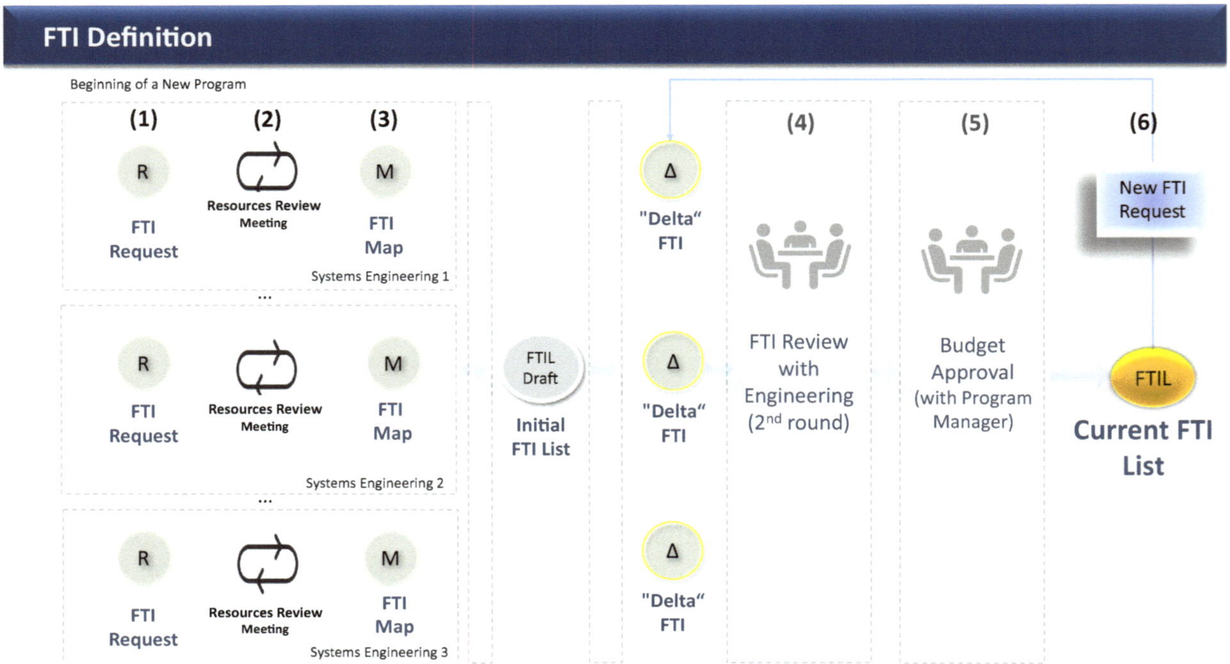

FTI Definition Process

TYPES OF PARAMETERS

- Analog data (position, force, angle, acceleration, pressure, temperature, strain, and flow sensors).
- Digital bus data (ARINC 429, MIL1553, RS-422, etc.).
- Video cameras (usually digital).
- High-speed cameras (above 100Hz and up to 1kHz).

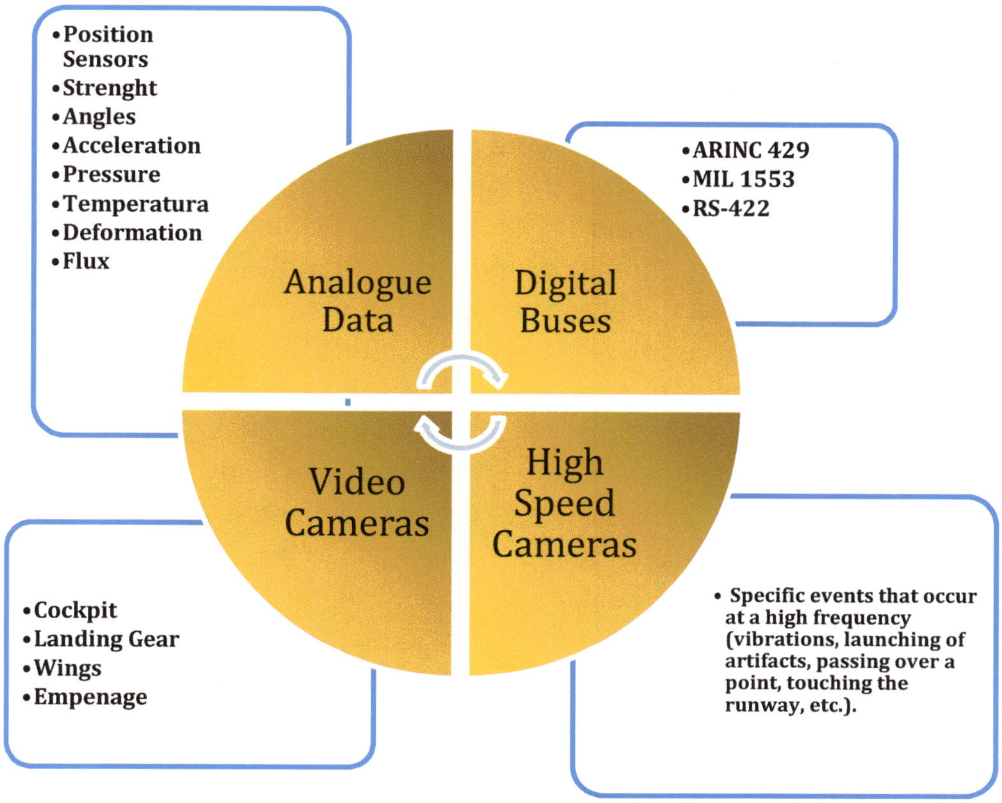

Definition of Flight Test Instrumentation

Each parameter needs a detailed definition:
- Sensor type (as listed above).
- Measurement range: minimum and maximum values that the sensor must be able to measure.
- Precision: closeness of the results to each other, related to the relative consistency of the results.
- Accuracy: closeness of measurements to the true value, related to absolute correctness.
- Sampling rate.

Defining the FTI list requires a lot of interaction between the FTI team, Flight Test Engineering, and Systems Engineering. This list, once defined, must be reviewed to ensure that all the parameters are included and well-defined. However, it is common for new parameters to be requested to assess phenomena in more detail or increase the safety of tests.

It should also be noted that each Test Campaign has a dedicated list of FTIs, a subset of the complete list described in the previous paragraph. This list must be defined for each Flight Test Plan (FTP). Therefore, throughout an aircraft development campaign, the FTI is constantly being changed to meet the specific needs of each test campaign.

FTI systems usually comprise the following parts:

1. Measuring sensors: convert a physical quantity into a measurable electrical signal.
2. Data Acquisition Unit: this is the heart of the system, responsible for collecting, conditioning, and digitizing the signals coming from the sensors.
3. Data Recorders: responsible for recording the acquired data for later analysis.
4. Telemetry systems: transmit and receive data in real time for analysis during tests, enabling data analysis and decision-making in real time with the support of a team on the ground.
5. On-board computer: receives and processes data for display by the Data Analysis software during the flight.
6. Data Analysis Software: makes it possible to display and analyze data. Usually, the same software is also used in the aircraft and telemetry stations. This allows complex calculations to be made and limits to be monitored, as well as providing feedback to pilots in real time.
7. Wiring and Connections: needed to connect the sensors to the Data Acquisition Unit and other FTI components.

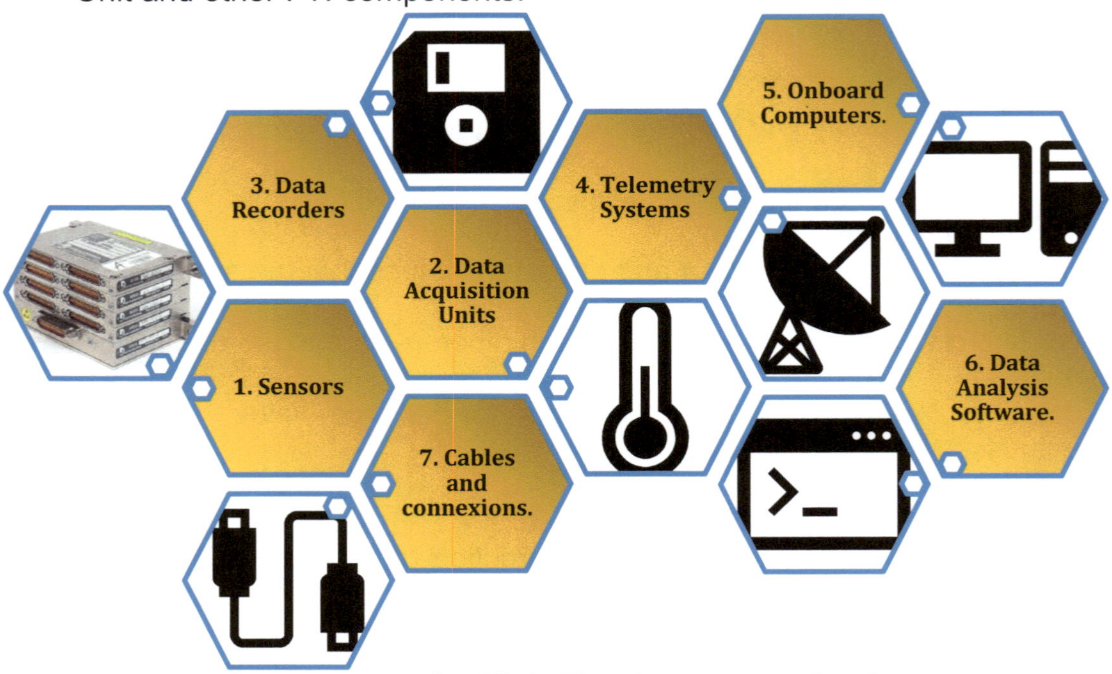

Components of a Flight Test Instrumentation System

Systems Engineering does aeronautical design but does not operate the systems it designs. Test Engineering and Test Pilots do not design but operate the systems they design. FTI is an example of a system that does both design and operation (of the FTI system). That is why its members must be able to do both, and usually concurrently. This requires specialized professionals with a great deal of experience.

The FTI parameter measurement chain must ensure that the design criteria are met (measurement range, sampling rate, precision, accuracy).

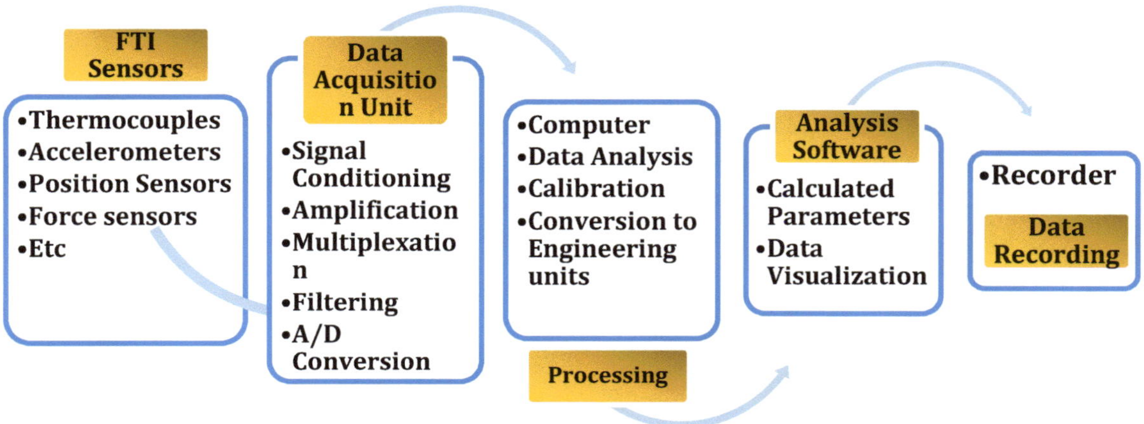

FTI Measurement Chain

SENSOR CALIBRATION

The correct reading of parameters acquired by sensors requires them to be calibrated and a calibration curve to be in place. Often, the company does not have the equipment and standards to carry out these calibrations. This service must then be subcontracted. It is usually worth having an FTI calibration laboratory within the organization for financial reasons, but it depends on the existence of a trained team within the organization.

Certification requirements demand that the FTI has been calibrated for the results to be accepted. This calibration is normally valid for one year.

FTI MANAGEMENT

As mentioned earlier, the FTI must be constantly adjusted throughout a Flight Test Campaign. It needs to be installed even during the production phase of a new aircraft because some sensors and wiring depend on access that is easily available only during production. It usually remains installed on prototype aircraft throughout their life, or until the prototype has to be delivered to a customer.

Throughout the long period, the FTI is installed, it needs to be managed similarly to other aircraft systems.

The initial project, which includes all the tests that will be carried out, often undergoes adjustments that require new FTI parameters. These new parameters are design modifications that require new designs, components, and work instructions to be installed, checked, calibrated, and made available for analysis.

FTI and aircraft changes should be managed in periodic configuration meetings involving representatives of the areas involved in Prototype Configuration Control (see Chapter 3, The Flight Test Process).

A widely used form of FTI structure is shown below. In some Flight Test Organizations, the Database Area is separated from the FTI for reasons of professional specialization and the need for other types of resources.

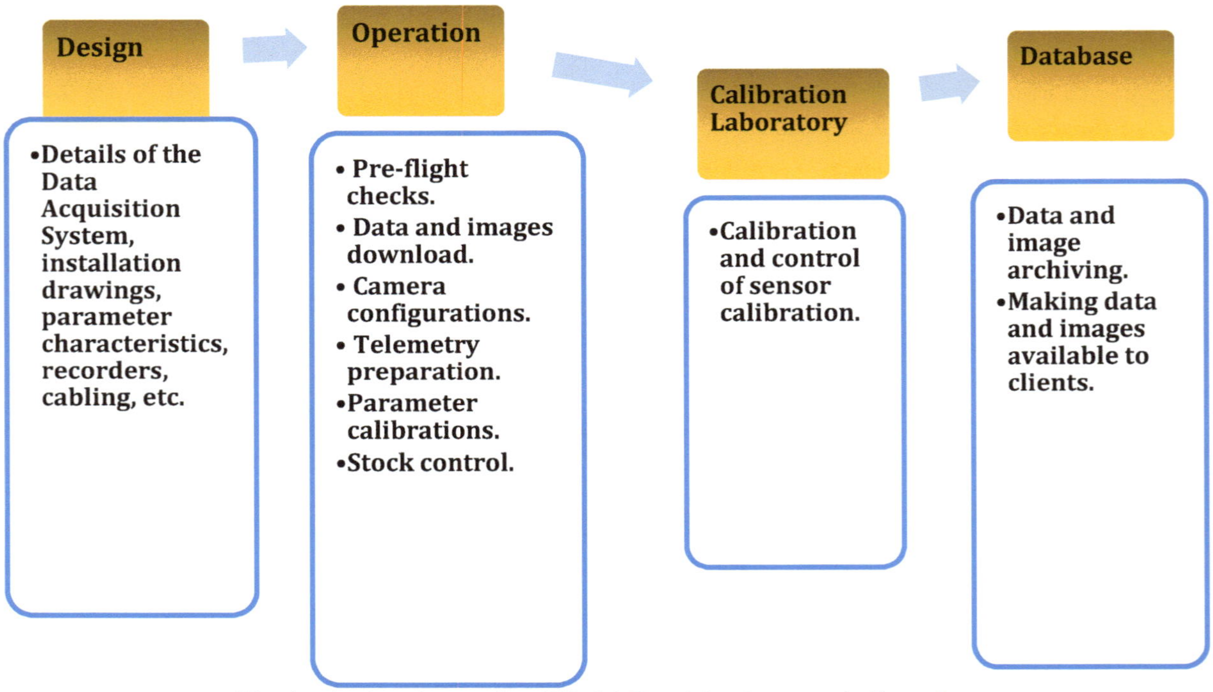

Typical structure of a Flight Test Instrumentation Area

FTI BASELINE AND DELTAS FTI

The FTI Baseline is a concept that helps define the FTI for the first aircraft tests. It consists of a set of initial parameters required for the first flights. These flights include the First Flight and the Functional Check flights. A Baseline FTI should be defined because functional check flights are usually repeated, in part or in full, after modifications have been made to the aircraft.

It usually contains many parameters related to Flight Mechanics and Systems (altitude, speeds, fuel quantity, latitude, longitude, heading, roll angle, pitch attitude, angle of attack and slip, flight control deflections, flap and compensator deflections, engine parameters, basic electrical, hydraulic, environmental control and navigation system parameters, video cameras, outside air temperature, etc.). For fly-by-wire aircraft, the FTI Baseline is much more complex, as it requires digital flight control parameters.

As the Test Campaign evolves, new parameters may become necessary. These new parameters are installed in small packages called Deltas FTIs.

However, we must not confuse the Delta FTIs with the new FTI configurations needed to support Test Campaigns (see figure below).
FTI configuration changes should be planned in the Long-Term Meetings and coordinated in the Short-Term Meetings (see Chapter 3, The Flight Test Processes).

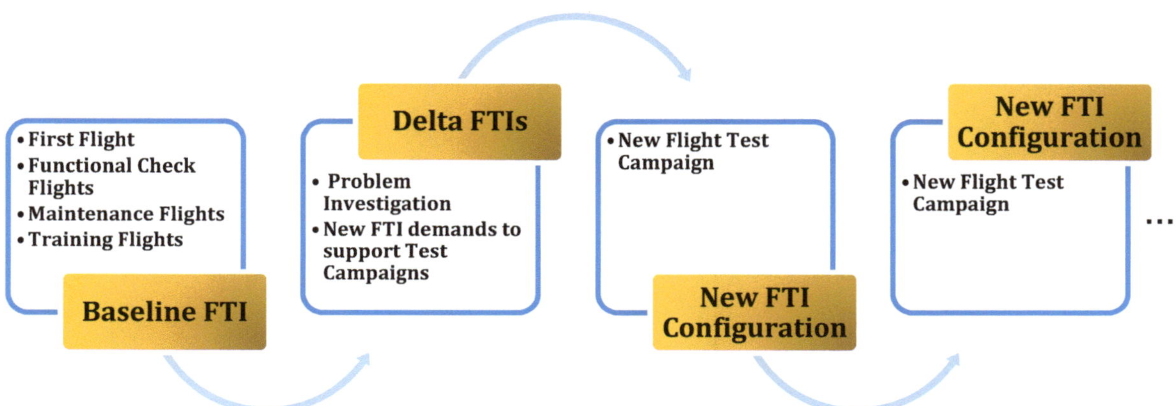

FTI Baseline, Delta FTIs, and New FTI Configurations

FLIGHT TEST MAINTENANCE

9

"Action solves anything."
Author unknown

AIRCRAFT MAINTENANCE

Prototype aircraft maintenance is a particular case of maintenance for any aircraft. However, it is a case in which many of the maintenance procedures are not yet well defined or simply do not exist. This refers to:
- Maintenance tasks
- Inspection intervals

Before we talk about Flight Test Maintenance, we must first discuss Aircraft Maintenance in general.

An aircraft is maintained following the guidelines in the Aircraft Maintenance Planning Document (MPD) and the Aircraft Maintenance Manual. An aircraft's Maintenance Plan is the result of efforts on different fronts using different methods.

The main sources of an aircraft's Maintenance Plan are:
- System Safety Assessments (SSA) done by the aircraft manufacturer. This method is described in Advisory Circular AC-1309 and AC 25-19A.
- The report produced by the Maintenance Review Board Report (MRBR), if the manufacturer uses the Maintenance Steering Group MSG-3 method.

These two methods are combined to generate a certifiable result, and at the same time embrace the best maintenance practices accumulated over decades in the MSG-3.

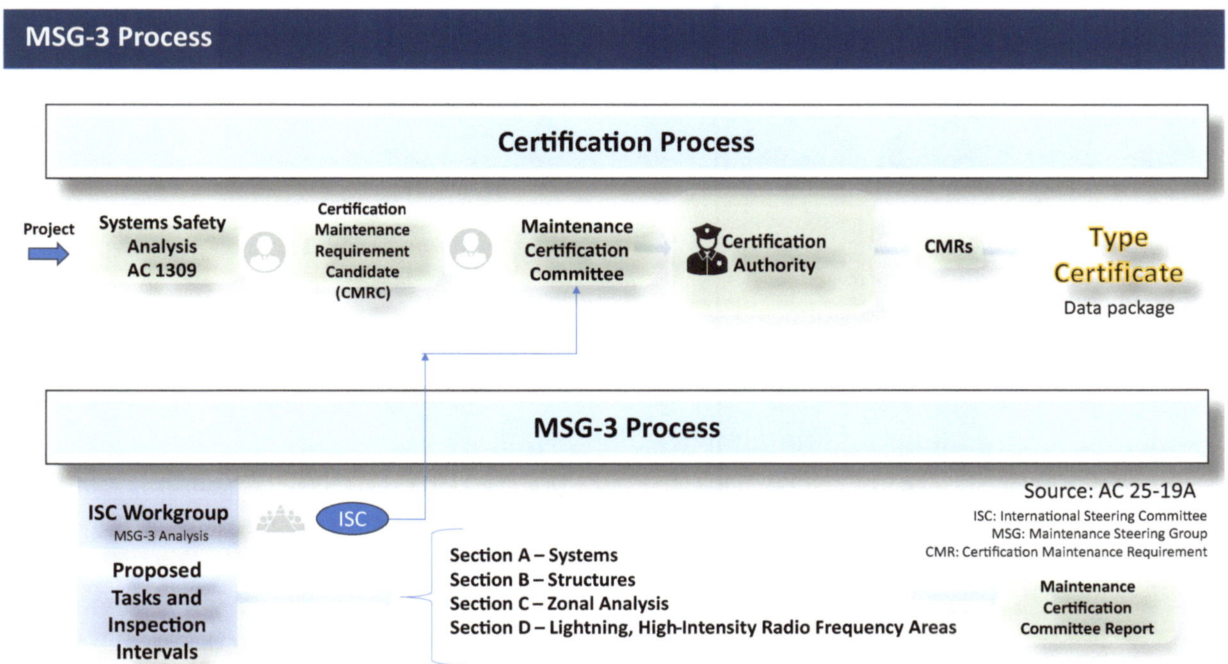

Development of Scheduled Maintenance Tasks

THE MSG-3 MAINTENANCE STEERING GROUP METHOD

The MSG-3 method is the most widely used method for developing a maintenance program and inspection intervals. It is an evolution of the MSG-1 and MSG-2 and is accepted by aviation authorities around the world. Its latest edition was approved in 2005.

MSG-3 is a program focused on tasks. Tasks are designed to protect systems from possible failures.
One or more tasks may be required depending on the characteristics of the fault.

MSG-3 analysis consists of four parts:
- Systems Analysis
- Structural Analysis
- Zonal Analysis
- Lightning and High-Intensity Radio Frequency Areas

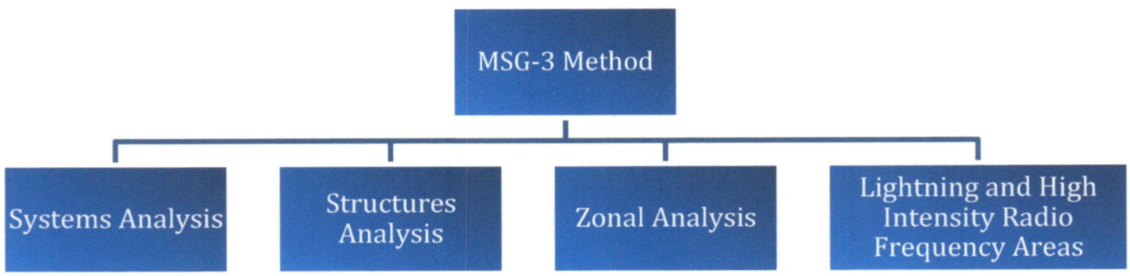

MSG-3 method

The MSG-3 method classifies maintenance tasks into five types:

1. Lubrication/Supply (LU/SV or LUB/SVC)
2. Operational/Visual Check (OP/VC or OPC/VCK)
3. Inspection or Functional Check
    a. - General Visual Inspection (GVI): carried out to detect obvious defects or irregularities in components. It does not use any special tools and is usually carried out in natural light.
    b. - General Detailed Inspection (GDI): intensive inspection of a specific component to detect defects or irregularities. It usually requires artificial light and sometimes the opening of doorways. The use of magnifying glasses or mirrors and surface cleaning of the parts are also usually required.
    c. - Special Detailed Inspection (SDI): inspection of a component using special tools such as non-destructive inspection (NDT) techniques, borescope or video-borescope inspections. It usually requires disassembly and thorough cleaning of the components.

4. Restoration
5. Disposal

SYSTEMS ANALYSIS

Systems analysis follows a top-down approach from each ATA, sub-ATA, and component. This analysis is done in five parts:

1) Identification of the candidates for Significant Maintenance Items (SMI).
SMI is identified in six stages:

a) Firstly, the aircraft is divided into large functional areas: ATAs, and sub-ATAs.

b) Top-down systems analysis is carried out to identify possible MSI candidates. Candidates will be those subsystems or components whose failure will impact safety, operation, or the cost of operations.

c) A list of questions is made for each MSI identified in the previous step:
- Is the failure of this system detectable?
- Does the failure of this system affect security?
- Does the failure of this system have a significant operational impact?
- Does the failure of this system have a significant economic impact?

d) For each item that had a "Yes" answer in the previous step, analysis must be carried out. An MSI is usually a system or subsystem. This is why this analysis must be started at the appropriate level so that it is not too high to make the analysis too complicated, but also not too low so that all functionalities and failure cases are analyzed. This level is called the highest manageable level.

e) Once the highest manageable level has been determined, the MSI list is sent by the manufacturer to the Industry Steering Committee (ISC). The ISC, in turn, reviews and sends the MSI list to the Working Groups.

f) In the sixth stage, the list of MSI candidates is analyzed by the "Working Groups" according to the MSG-3 methodology.

The failure of protection systems must also be taken into account. Protection systems are those that:
- Draw the crew's attention to abnormal functioning.
- Shut down equipment in the event of a fault.
- Eliminate, or reduce, the effect of a fault.
- Take over the function of failed equipment.

2) Analysis
MSG-3 analysis uses a logic diagram similar to the one below. The analysis is carried out on two levels.

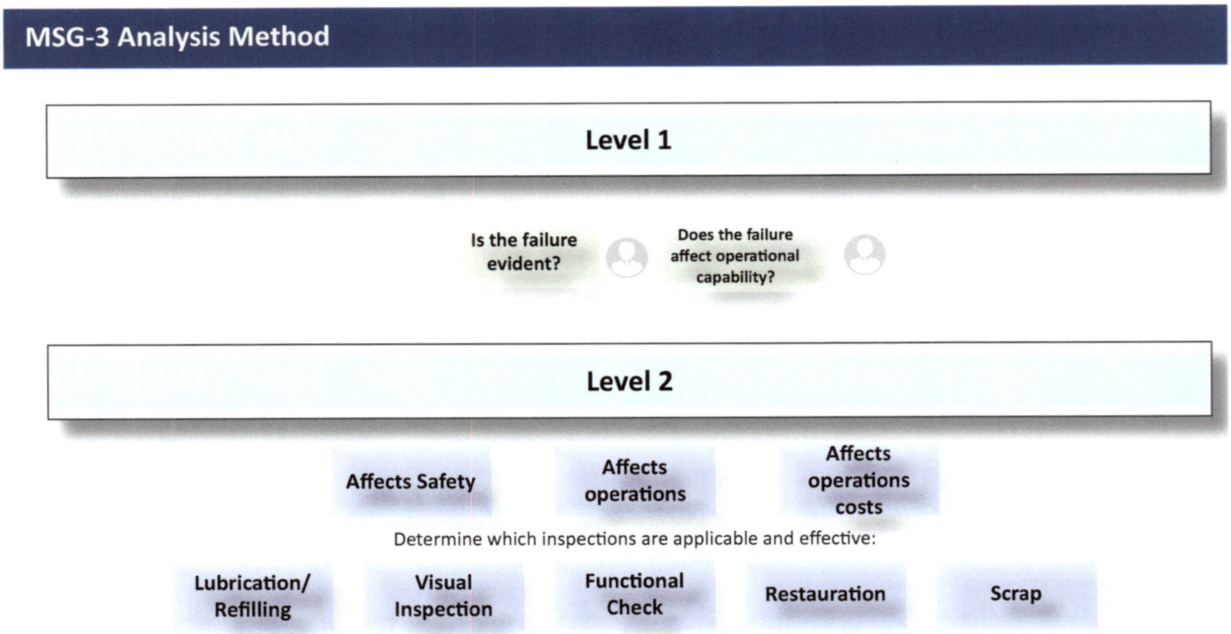

MSG-3 Analysis Method

LEVEL 1 ANALYSIS

Level 1 analysis is a study of the failure and its effects, whether the failure is announced to the crew, and, if not, how difficult it is to identify. Failures are classified into five categories of effects:

- Evident faults that affect safety (Category 5).
- Evident faults that affect operation (Category 6)
- Evident faults that affect the cost of operation (Category 7).
- Non-evident faults that affect safety (Category 8).
- Non-evident faults that do not affect safety (Category 9).

For each category, a logical sequence of questions was defined to determine the types of tasks required.

LEVEL 2 ANALYSIS

At this stage, the causes of the failure must be determined and suitable tasks selected to mitigate the effect of the failure. When the risk cannot be reduced by a task or combination of tasks, the system must be redesigned.

5) Determining Inspection Intervals

The final step is to determine the inspection intervals. These intervals can be defined according to the calendar (days, months, or years), or according to the life of the component. The criteria used to define the intervals must take into account:

- Consequences of failure
- Possible fault detection
- System or component reliability
- Engineering judgment
- Supplier or data operator

MSG-3 Method Systems Analysis

STRUCTURAL ANALYSIS

Like the work done on the systems side, Structure Analysis seeks to determine the Structure Significant Items (SSI). SSI are any parts of the structure, a single piece or assembly, that support loads and would compromise the aircraft structure if they were to fail.

Structural analysis must take into account corrosion, damage that may occur during operation, fatigue, and the environmental conditions the structure will be subjected to. For the sake of understanding, aircraft structures are divided into:
- Primary Structures
- Other structures

Once the SSIs have been identified, they are classified according to a scale. This scale is used to assign inspection intervals according to this classification and the effects of each SSI.

Structural analysis also needs to consider damage caused by accidents, and environmental damage.

Damage caused by accidents must consider the type of damage, the probability of the accident, and the probability of detecting the damage.

Damage caused by environmental factors must consider exposure to environmental damage, and the probability of detecting the damage.

THE MAINTENANCE CERTIFICATION METHOD

According to AC 25-19A, the System Safety Assessment method should identify Certification Maintenance Requirement Candidates (CMRC). These requirements are those maintenance tasks that can detect latent faults that in combination with specific faults can lead to a Hazardous or Catastrophic events.

Tasks can also be selected to determine components that are worn out by use.

The Certification Maintenance Requirements (CMR) must specify the failure mode to be detected, the failure condition, the type of inspection, and the inspection interval. All latent faults must be assessed.

It is important to note that AC 1309 recommends that quantitative analyses complement but do not replace engineering judgment based on justified experience, and, therefore, CMRs can be created based on this judgment.

As AC 1309 also regulates Major events, if a latent fault in combination with more specific faults can lead to a Major event, it may require a CMR (but such cases are uncommon).

THE MAINTENANCE CERTIFICATION COMMITTEE

The Certification Maintenance Coordination Committee (CMCC) brings together representatives of manufacturers, Certification Authorities, the Maintenance Review Board (when the MSG-3 method is used), and aircraft operators. In this way, the certification maintenance requirements are analyzed together with the results of the MSG-3 Maintenance Review Board (MRB) analysis.

This committee should be formed as early as possible during a certification program and meet regularly to review the Certification Maintenance Requirements (CMR) Candidates. This review should include the criticality of failures, the maintenance tasks, and other relevant aspects of the tasks.

In the end, the Certification Authority must approve the list of CMRs that will form part of the limitations of the Aircraft Type Certificate.

The manufacturer must send the list of CMRs to the Industry Steering Committee for consideration.

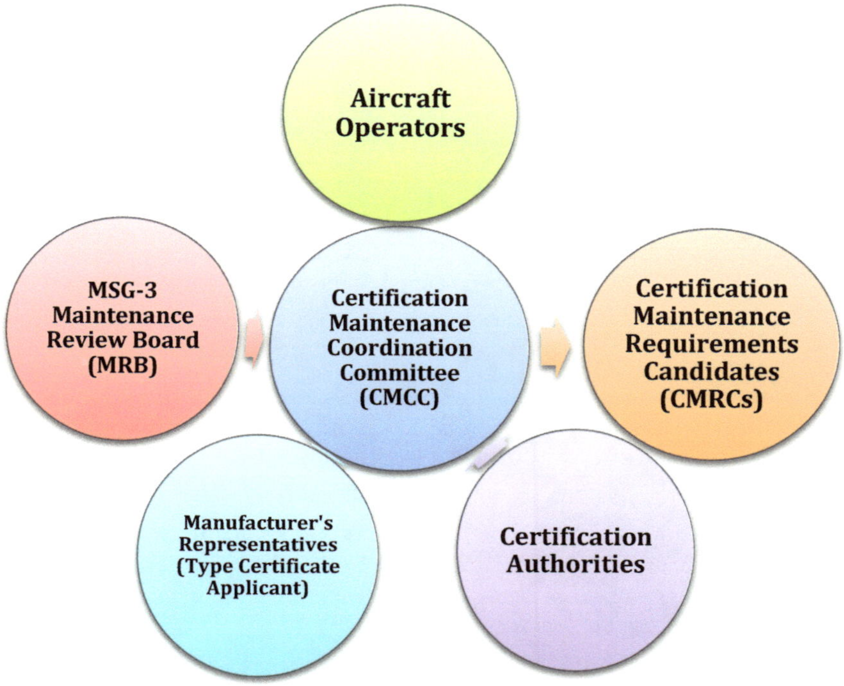

Certification Maintenance Coordination Committee (CMCC)

AIRCRAFT MAINTENANCE PLAN

An aircraft Maintenance Plan must contain:
- Tasks to be completed
- Description of tasks
- Man-hours required to carry out the tasks
- Task code:
    - BSI: borescope inspection
    - GVI: Visual Inspection
    - CHK: check (for leaks, quantity of fluids, accumulator pressure, etc.)
    - VC: Visual check
    - DI: Detailed Inspection
    - DS: Discard
    - FC: Functional Check
    - LU: Lubrication
    - OP: Operational Check
    - RS: Remove to Restore
    - SDI: Special Detailed Inspection
    - SV: Refilling fluids ("servicing")
    - TPS: Temporary Protection System

- Required Skill Code:
    - electrical system
    - cabin
    - motor
    - non-destructive testing
    - avionics
    - structure
    - radios
    - utilities

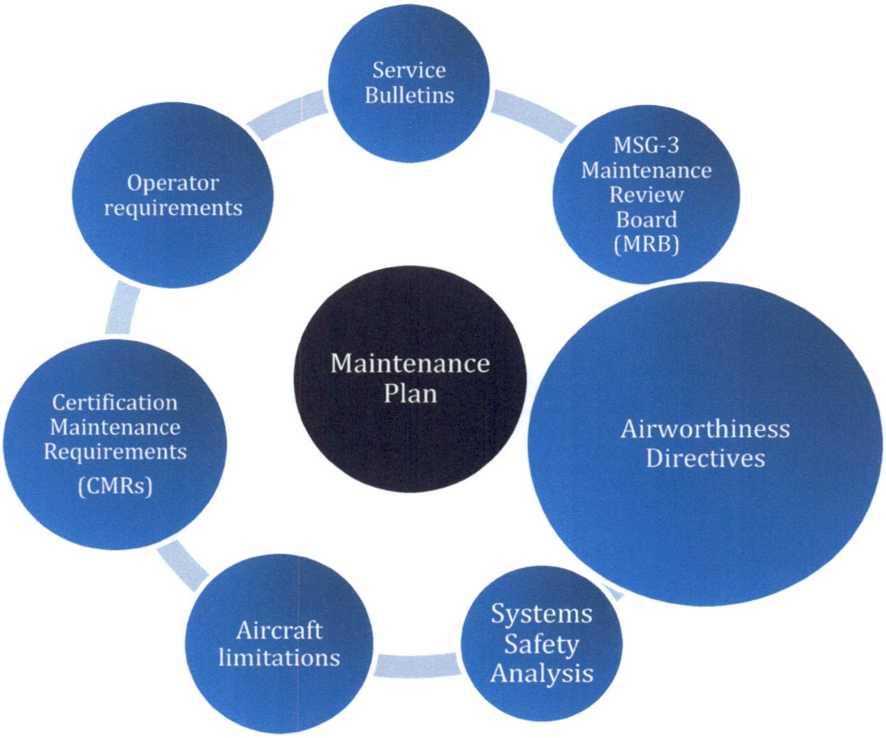

Maintenance Plan

The Aircraft Maintenance Plan should also contain a graph showing how these activities are distributed over time, as below.

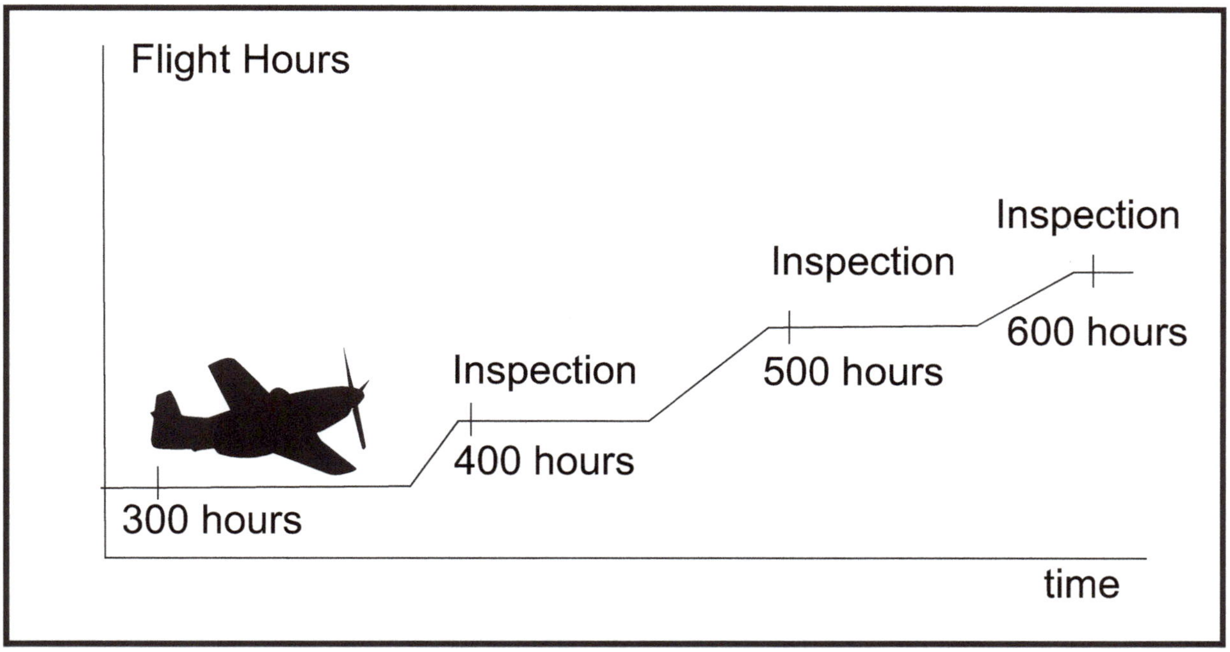

Scheduled Inspections

In addition, the Maintenance Plan should also contain another graph showing how the demands in man-hours are distributed over time (see figure below).

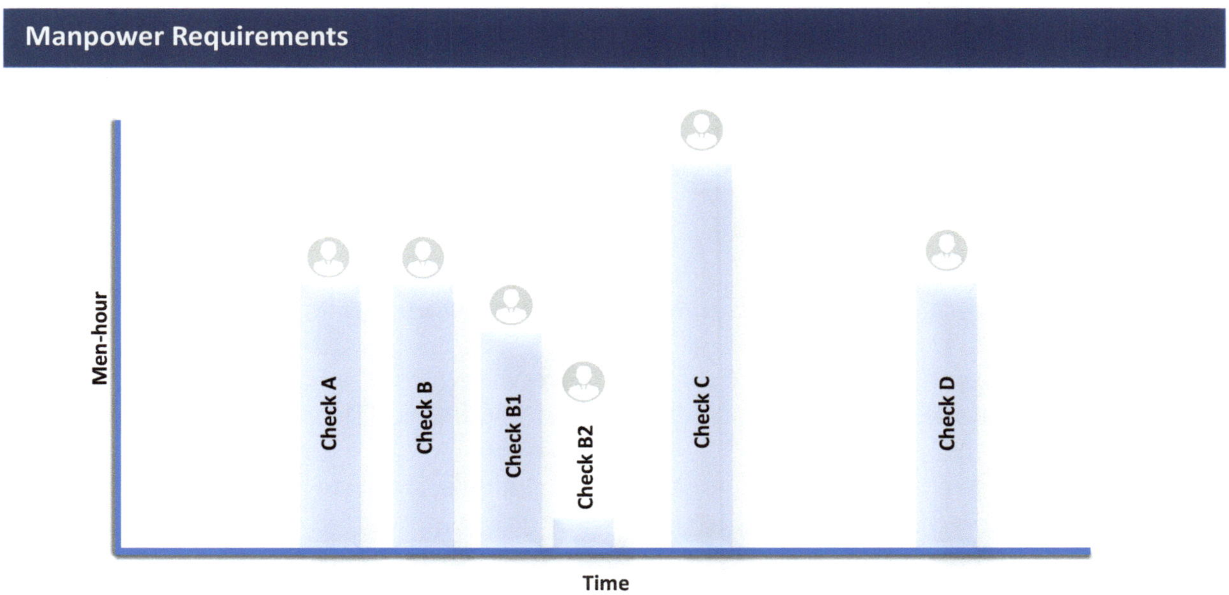

Distribution of man-hours over time

If we look at the distribution of man-hours in the figure above, we can see that it is very uneven. This makes it very difficult to organize work logically. The result is that the workforce is idle for long periods and extremely busy at other times, which means that, in addition to the loss of idle hours, there is a need for greater expenditure (more working hours) at peak times.

It should be remembered that the purpose of Maintenance is to deliver a service that guarantees flight safety, meets operational requirements, and is economically efficient for the organization's business.

For this reason, best maintenance practices call for equalizing the demand for working hours throughout the year to minimize peaks in demand. This is achieved by dividing large tasks into parts, so that they are distributed over several periods, thus avoiding peaks.

Equalized Maintenance Plan

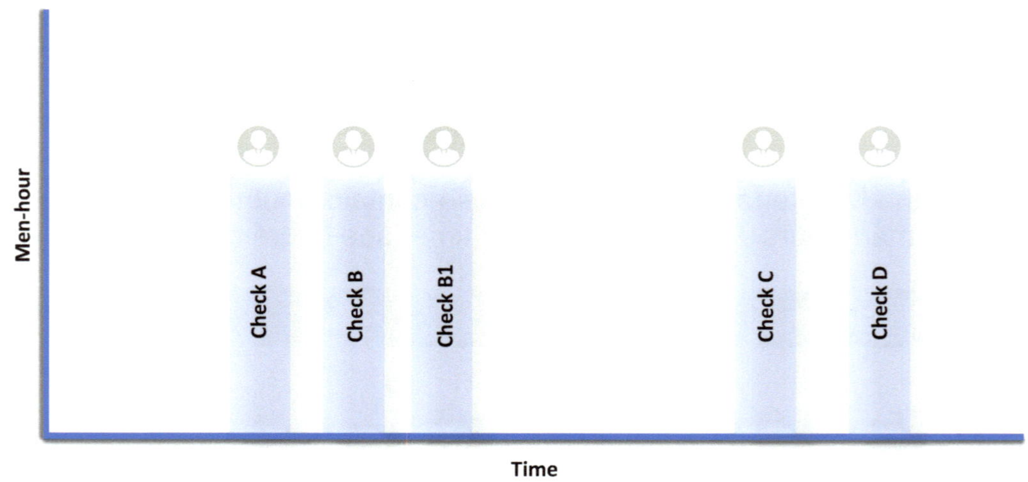

Equalized Maintenance Plan

In addition to the Maintenance Plan, there is another very important document for aircraft maintenance called the Maintenance Manual. The Maintenance Manual must contain:
- A detailed description of the maintenance activities required.
- A description of tools and other support equipment needed for maintenance.

However, this manual will often be written during the development of an aircraft, and therefore either incomplete or not available at that stage.

STRUCTURAL INSPECTIONS

An initial inspection is normally conducted followed by inspection performed at intervals.

Inspection intervals are defined according to:
- Flight hours
- Cycles or
- The calendar

An operator can convert the intervals based on the type of use of his aircraft, as long as they do not exceed the inspection intervals in the Maintenance Plan.

Training adjustments can be made considering that a complete landing or two touch-and-goes are equivalent to a complete cycle.

FUNCTIONAL AREAS OF FLIGHT TEST MAINTENANCE

To meet the maintenance requirements of all its customers, Flight Test Maintenance usually has a complex structure that requires an experienced team. One possible and efficient structure is shown in the next page.

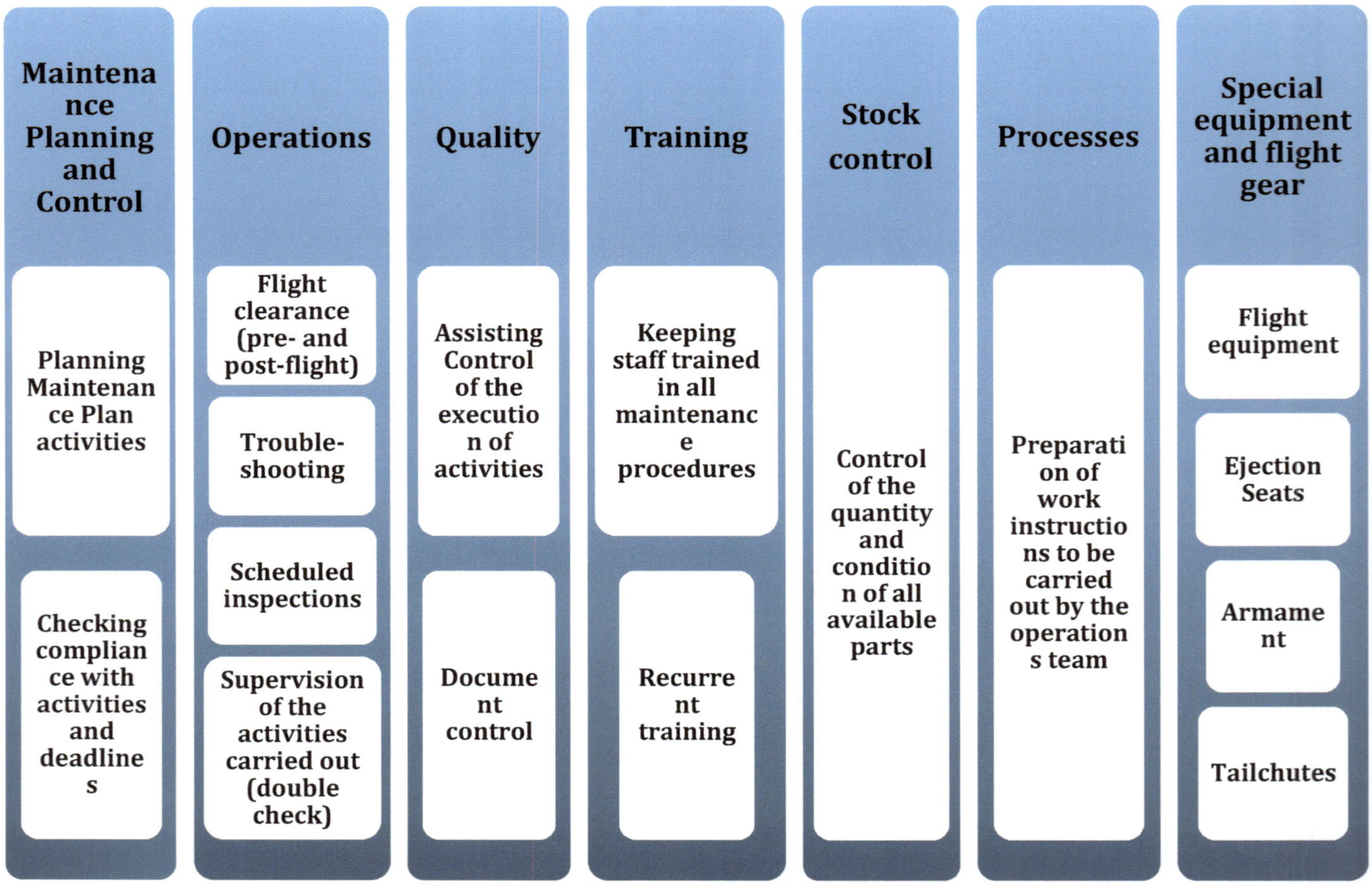

Typical structure of a Flight Test Maintenance Area

SAFETY MANAGEMENT SYSTEM (SMS)

10

"Expect the unexpected."
Popular saying

The Safety Management System (SMS) is a systematic, documented, and proactive process that seeks to improve the safety of an organization. It aims to create proactive activities to prevent accidents. As a principle of SMS, all company employees are responsible for measures aimed at improving safety. Ultimately, SMS aims to prevent accidents.

It is made up of four parts, or pillars, that complement and interrelate with each other. They are:

- Prevention Policy
- Promoting Safety
- Risk Management
- Safety Assurance

SMS Pillars

PREVENTION POLICY

These are actions that are the responsibility of the company's top management, and which aim to guarantee a commitment to safety. Prevention policies must guarantee an appropriate structure for safety throughout the organization. They also aim to clearly define the organization's safety objectives. Prevention policies must be documented in such a way as to guarantee the transparency of all security-related processes.

SAFETY PROMOTION

Promoting safety culture consists of disseminating the safety culture, its principles, results, and lessons learned. It also involves providing training when necessary.

RISK MANAGEMENT

Risk management consists of the process of identifying threats, analyzing threats, defining risks, and, finally, controlling risks through mitigation actions.

SAFETY ASSURANCE

The SMS process must be checked to make sure it is working as intended. Safety Assurance aims to guarantee the effectiveness of the mitigation actions taken during the Risk Management process and to identify new threats. It is a process that takes place during operations and uses facts and data to assess whether the results of operations are as expected.

Risk Management and Safety Assurance are the two most important processes in the SMS because they are the ones that effectively deal with each risk or threat.

ICAO Annex 19 requires Project or Production Organizations to have SMS. It should be noted that accident investigation is not part of SMS.

Despite its pillars, SMS depends heavily on a safety culture and the commitment of the organization. It recognizes that safety depends on processes that manage the causes of problems, not their effects.

Accidents are events that compromise organizations in many ways, with serious and lasting consequences. But above all, they generate:
- Human costs
- Material Costs
- Image costs
- Liabilities

However, it is a recognized fact that the total elimination of accidents is not attainable.

As a result of ICAO Annex 19, the FAA started the Part 21/SMS Rulemaking Project in 2014. With input from several manufacturers, the Aviation Rulemaking Committee (ARC) continued the project.
In addition, the National Aviation Standard (NAS) 9927 was developed by the American Aerospace Industries Association (AIA) and the General Aviation Manufacturers Association (GAMA).

The FAA encourages the aviation industry to implement an SMS and have it formally recognized by the FAA.

Important definitions related to SMS:

Threats
These are potential loss scenarios made up of three components: source, mechanism, and result.

Risks
These are a combination of something that could go wrong, the probability of it happening, and the consequence if it does.

It is common not to be sure what is a threat and what is a risk. A threat is what you see, a risk is the result of analysis.

Errors
These are unsafe actions that can cause unwanted consequences, such as accidents and incidents.

Violations
These are errors caused intentionally.

Human error
Estimates suggest that human error has become the main cause of accidents and incidents in the aviation industry, growing from 20% in the 1960s to around 80% in the 1990s. At the same time, errors due to material failures fell by an inverse proportion, from 80% to around 20%.

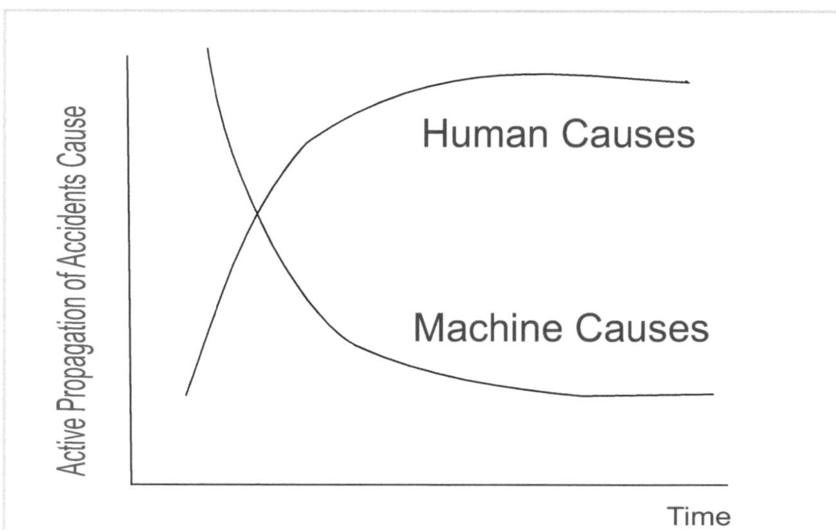

Cause of Accidents (Human and Machine Faults)

Due to the growing importance of human error in the occurrence of accidents, various models have been created to study it.

JAMES REASON'S THEORY (SWISS CHEESE)

Professor James Reason, from the University of Manchester, published his theory in the book *Human Error* in 1991. James Reason's theory has become an important theory for explaining the occurrence of accidents and incidents. It is used not only in aviation, but in other industries, and medicine. He argues that accidents in complex systems do not occur due to isolated failures. This theory is based on the pre-existence of failures in the organization. These failures can be of two types: active, or latent. Active failures are those that have immediate action. Latent failures are those that are often unknown and can take action long after they have been generated. This theory is best explained by Swiss cheese. The holes in the cheese are the faults, and the processes are represented by several aligned cheeses. When the faults of several cheeses line up, an undesirable consequence, or even an accident, can occur. When the holes in the cheeses are misaligned, it means that the system is protected by barriers or defenses against the occurrence of accidents and incidents.

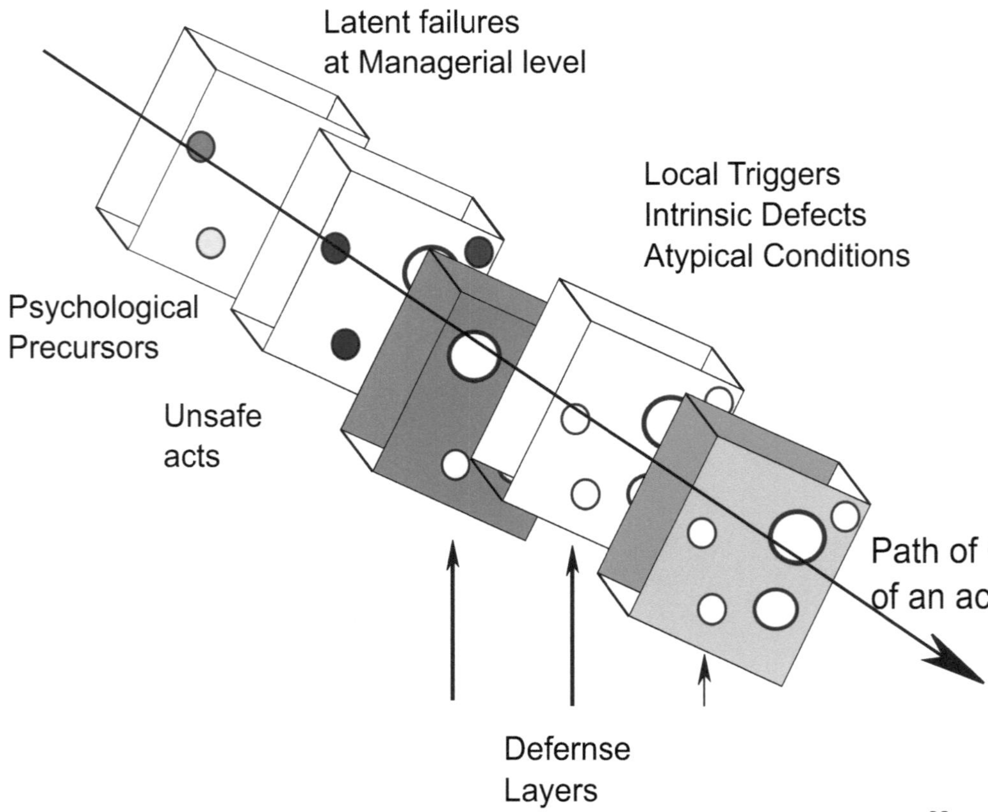

Principle of the Reason Model (Swiss Cheese)[22]

Another more detailed way of explaining Reason's theory relates Organizational Influences, Work Environment Factors, and, finally, unsafe acts (active failures) that cause accidents and incidents (see the following image).

[22] See Reference 16.

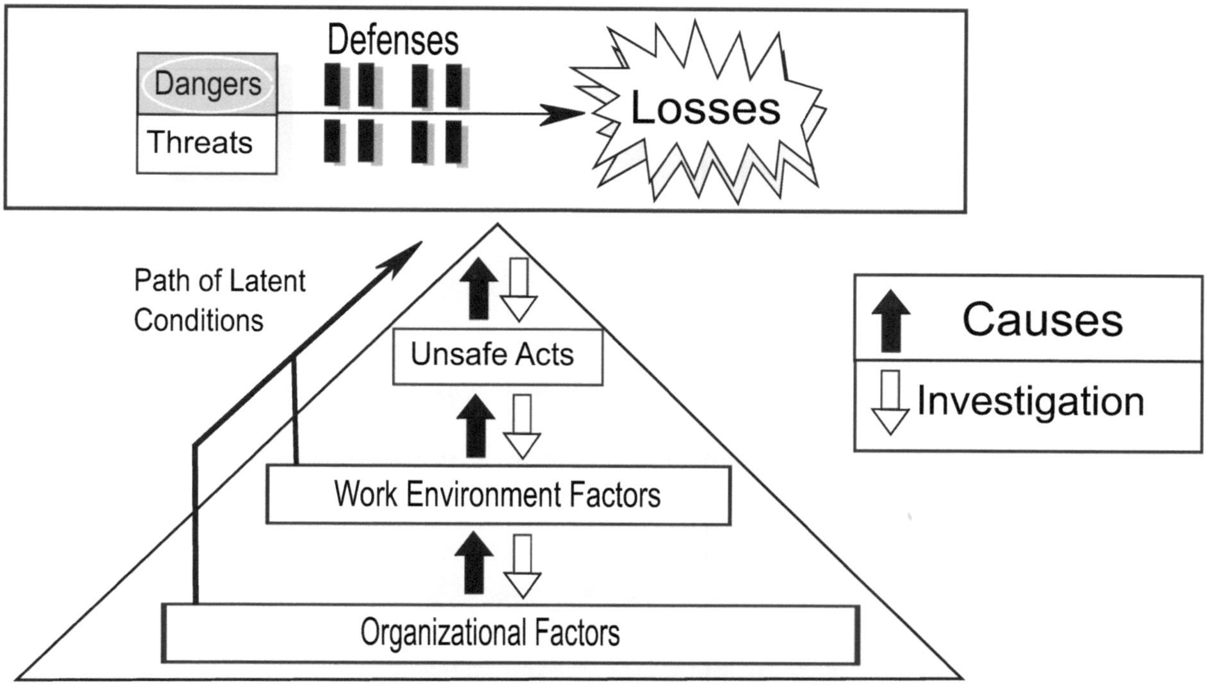

Reason model (Swiss Cheese)[23]

THE BOWTIE MODEL

The "bowtie" model seeks to graphically represent a failure mechanism. To do so, it places the possible causes on one side of the event and the possible consequences on the other. This method originated in the oil and gas industry in the 1990s and has spread to various industries, such as medicine and the aeronautical industry. It is currently recommended by the ICAO Safety Management Manual (SMM).
There is also software that uses the "bowtie" model and can serve as a basis for archiving and analyzing various occurrences.

There are barriers between threats and unsafe events. On the left, these barriers seek to prevent the event from occurring. On the right, these barriers seek to reduce the probability of the event occurring or to reduce the severity of its consequences.
The heart and value of the bow-tie method lies in the creation and management of barriers. The method can help answer questions, such as:

- What could go wrong?
- What barriers exist to prevent this from happening?
- Are these barriers working as expected?

[23] See Reference 16.

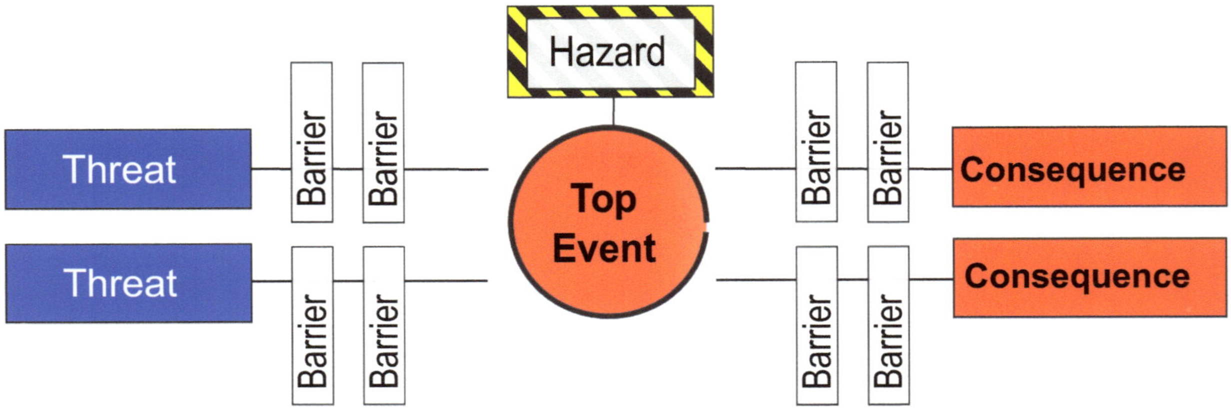

Bow-tie method

As an example, we can use the case of poor control of Weight and Center of Gravity (CG). Various barriers can be used to control this threat, as shown in the bow-tie diagram below.

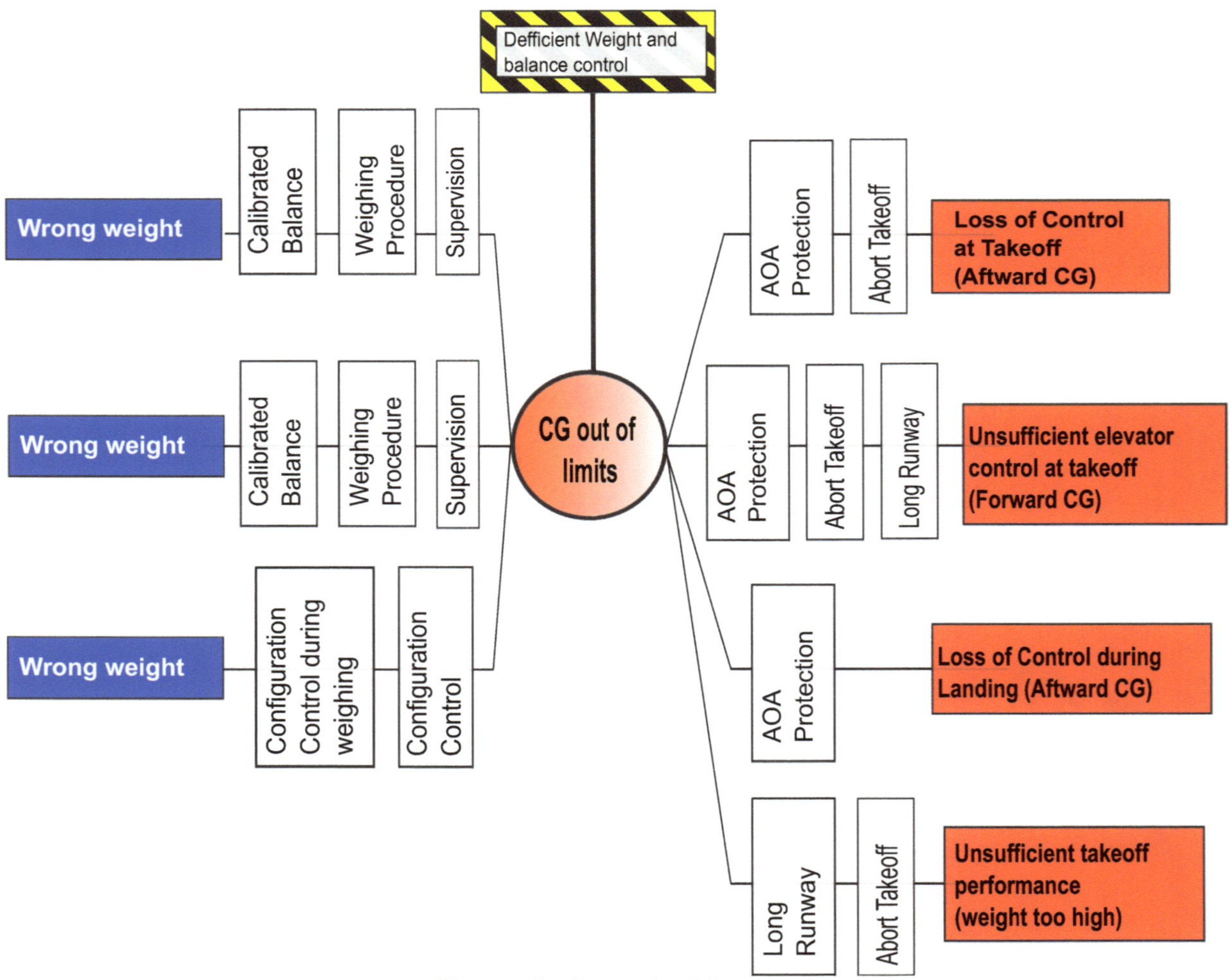

Example Bow-tie Diagram

The Bow-tie Method can be used both proactively to prevent accidents and reactively to analyze occurrences. This graphic representation is very useful when studying complex events, such as aircraft accidents.

THE SHELL MODEL (FOR HUMAN FACTORS)

The Shell model was developed by Frank Hawkins in 1975 and is described in ICAO document 9859. It considers latent conditions that exist in the system, as well as human factors, capabilities, and limitations when interacting with other system components. According to this model, a misalignment between the human being ("*liveware*") and the other three components ("*software*," "*hardware,*" and "*environment*") contributes to human error. So, these interactions must be studied and taken into account.

The Shell model

These interactions can be studied as follows:

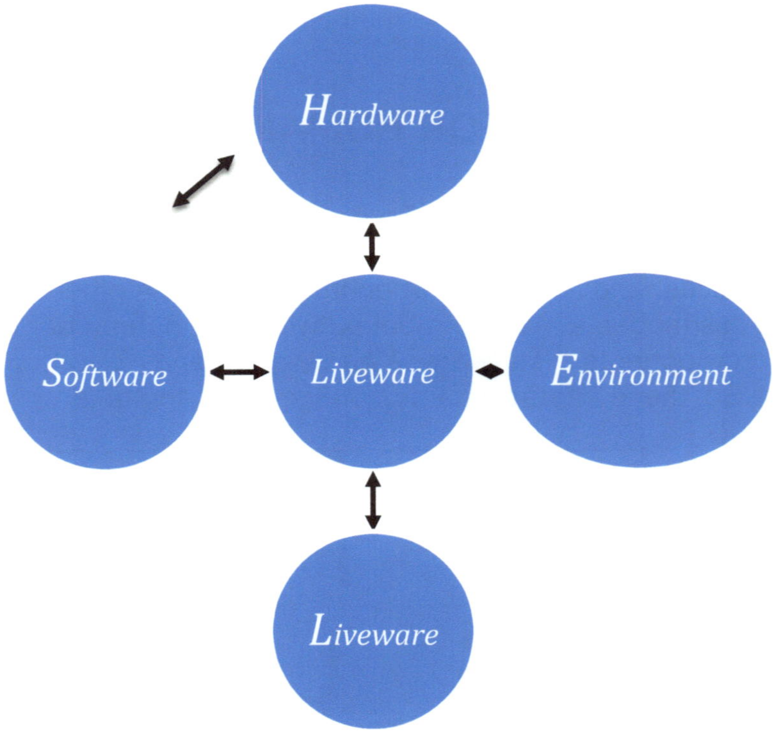

Interactions that occur according to the Shell Model

THE PEAR MODEL (FOR HUMAN FACTORS)

The Pear model was proposed by William Johnson and Michael Maddox in 2007. It considers that human errors are made up of:
- Physical factors
- Psychological factors
- Physiological factors
- Psychosocial factors

P	People
E	Environment
A	Actions
R	Resources

Pear Human Factors Model

ORGANIZATIONAL FACTORS AND RISKS

II

"Fear is a reaction. Courage is a decision."
Winston Churchill

ORGANIZATIONAL FACTORS

These conditions are far-reaching and very difficult to change. They affect programs from start to finish and are often defined at their inception. They become law and are often perpetuated in the organization. Correcting them is usually very costly.

Examples

- Treating suppliers like customers
- Contracts involving external organizations in internal processes
- Goals with inflexible or unattainable dates
- Heterogeneity of Business Maturity (Uneven IT Control Processes and Tools between different areas of the company).
- Lack of materials
- Lack of training
- Lack of human resources

ORGANIZATIONAL RISKS

They are the real symptoms of the Organizational Factors mentioned above. They are relatively easy to identify. Corrections are often procedural and inexpensive.

- Normalization of deviations (when the error becomes the standard)
- Documents that do not reflect reality, do not exist or are incomplete
- Interference by people from other areas in internal processes
- Little involvement from Development Engineering
- Lack of data available for engineering decisions
- Incomplete Production Tests
- Poor, or non-existent, performance by the Quality Area
- Latent faults on aircraft
- Repeated failures
- Constant complaints
- Relationship problems
- Deadlines set without considering all the important factors
- Resistance to follow existing standards

NORMALIZATION OF DEVIATION

The normalization of deviations can lead to very damaging consequences for the company. Avoiding the normalization of deviation:

1. Recognize your vulnerability.
2. Plan the flight and fly the plan (execute to achieve standards).
3. Consider your instincts (they exist for a reason).
4. Re-evaluate the lessons of the past.

THREATS TO OPERATIONAL SAFETY

Threats are a daily occurrence in aircraft development. But what are threats? They are conditions or situations that could potentially have serious consequences for people, aircraft, or the Development Program.

The threats are often clear, but sometimes it is hard to see them. Each set of tests also brings with it a set of threats. However, two things help us deal with threats and achieve our goals.

The first is the knowledge gained through successive tests. This knowledge often shows that certain threats were not actually important. That is why the threat set after testing is often smaller than before testing when some of these threats were better understood. For example, initially, anemometric errors are very poorly understood, and this necessarily creates the threat of an inadvertent stall. Many hypotheses are sometimes created to estimate these errors. However, once the anemometric calibration tests have been carried out, these errors are carefully measured and known, eliminating this threat, or making it unimportant.

The second factor to consider is design changes. During the development of a new aircraft, several changes are necessary. These changes aim to improve the performance of systems, correct deficiencies, or make new instrumentation measurements. Over time, these changes improve the design, and therefore reduce, or eliminate, threats.

These two factors above are the reasons threats are gradually reduced until they eventually reach a stage where they meet the level of maturity required by certification authorities and by the costumers.

There are always exceptions. These exceptions occur when a certain threat is covered up by the lack of knowledge that exists in certain operations. When this happens, the threats after testing increase because some of them have been revealed by the tests. These types of threats are dangerous and can lead to serious consequences. They are called dormant threats.

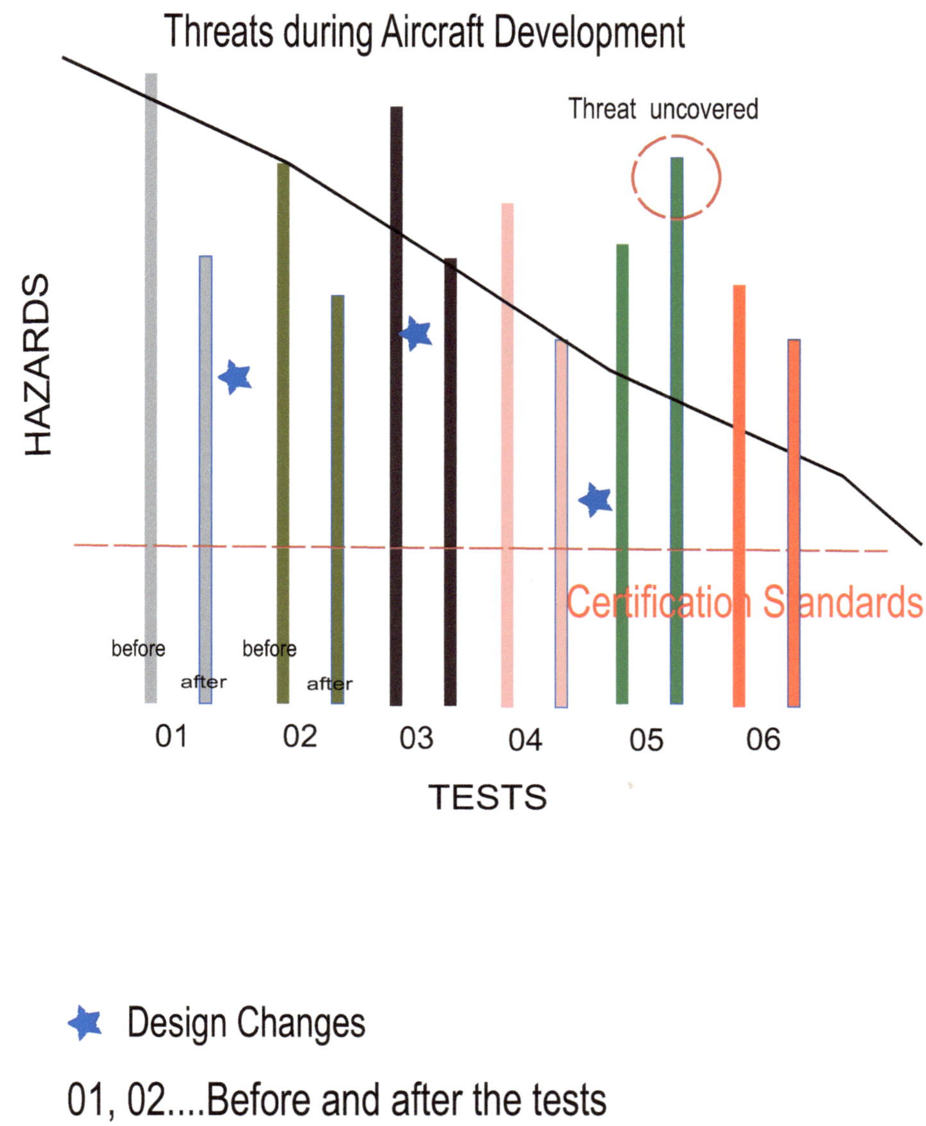

Design changes and knowledge gain over tests results help reduce hazards.

Threats during aircraft development

RISK MANAGEMENT IN FLIGHT TESTING

12

"It is very simple to make it well. But it is very difficult to be simple."

FLIGHT TEST RISK MANAGEMENT AND THE SMS

Flight Test Risk Management follows the same principles as SMS. It consists of identifying risks, quantifying or qualifying the degree of risk that the event will cause to people and resources, and taking actions to mitigate these risks. The risk assessment of a test campaign is always carried out by a test pilot and a flight test engineer.

However, it is important to differentiate SMS from Flight Test Risk Management. The SMS has its own standards and operating methods (The SMS Four Pillars) that differ from Flight Test Risk Management. It aims for comprehensive control of all threats to the company, its employees, and its facilities. Flight Test Risk Management, on the other hand, focuses on the events that are planned to occur during test flights, their particularities, techniques, and mitigations. There is often confusion between SMS and Flight Test Risk Management.

Flight Tests Risk Management identifies flight test threats, analyzes and mitigates risks, and, finally, classifies these risks.

The Flight Test Risk Management process takes place formally in two types of meetings:
- Flight Test Safety Reviews (FTSR)
- Flight Safety Committee Meetings (FSC)

The aim is to ensure that every threat (technical, maintenance, infrastructure, and operational) is identified and the risks analyzed, classified, mitigated, and documented before the test flights take place.

The Flight Test Safety Review ensures that all mitigation actions are defined before the tests are carried out. It shall be carried out before the Flight Test Plans are approved, to ensure that threats are taken into consideration by the plan.

The Flight Test team and Systems Engineering must be at the FTSR because the conclusions may include, modify, or exclude tests that have been requested by Engineering.

The Flight Safety Committee Meetings (FSC) aims to ensure that the aircraft is safe to be operated. It is not, therefore, required for every test plan. When a change or a result affects the airworthiness of the aircraft, an FSC should be convened to define actions to ensure that the aircraft is once again airworthy before testing resumes.

The Airworthiness, Flight Test, and Systems Engineering teams must be invited to the FSCM because its conclusions may approve or cancel the airworthiness of the aircraft.

The FTSR and FSC are approved according to the highest remaining risk level.

Mitigation actions are relevant information for the preparation and conduct of flights that will be discussed later during pre-flight briefings.

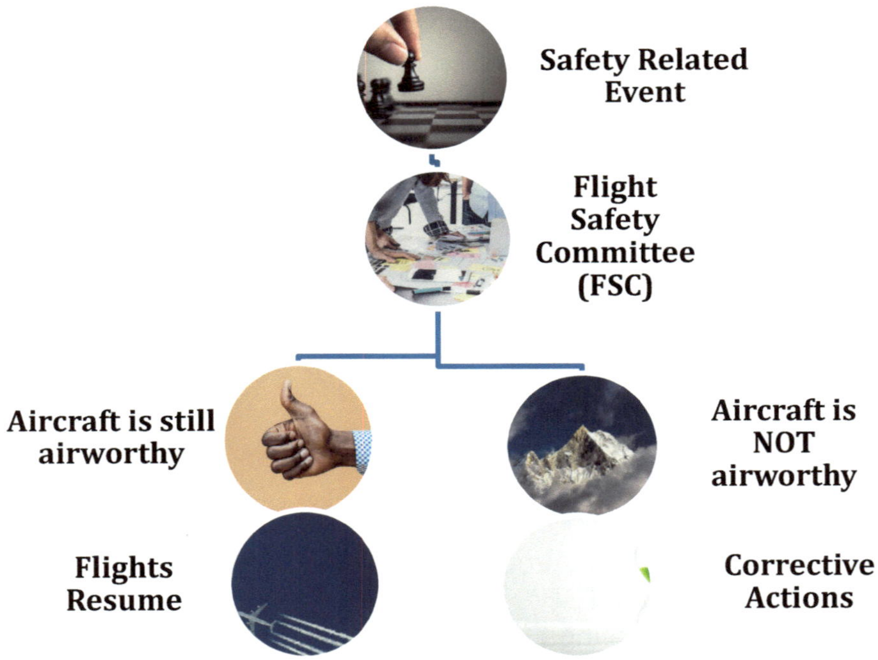

Safety Management System (SMS)

EVENTS THAT AFFECT SAFETY

In the aviation industry, safety-related events are classified into five categories: catastrophic, hazardous, major, minor, and non-safety significant events.

CATASTROPHIC EVENTS

These are occurrences related to the operation of the aircraft before landing, during which at least one of the following situations occurs:

- Someone suffers serious injury or death because of their presence in the aircraft, in direct contact with one of its parts, including insulated parts, or they are subjected to direct exposure to the propeller, rotor, or exhaust jet, or its consequences.

- The aircraft has suffered damage or failure that impairs its structural strength, performance, or flight characteristics and requires the replacement of major components or major repairs to the affected component. The exception is damage limited to the engine, its armor or accessories, or damage limited to the propellers, wing tips, antennas, tires, brakes, fairings, landing gear, light dents, and small holes in the aircraft's covering.

- The aircraft is missing or in an inaccessible place.

HAZARDOUS EVENTS

These are defined as all occurrences, including air traffic, associated with the operation of an aircraft that are not characterized as an accident as defined above, but which seriously affect the safety of operation.

MAJOR EVENTS

These are defined as failures that significantly reduce the safety margins or functional capabilities of the aircraft, significantly increase workload, or have an impact on crew efficiency.

MINOR EVENTS

These are defined as failures that present small reductions in flight safety and the functional capabilities of the aircraft. They require crew actions that are within their normal operating capacity or a small increase in workload.

NON-SIGNIFICANT EVENTS

These do not affect the aircraft's operational capacity or the crew's workload.

RISKS

The combination of probability and severity of the consequences of a hazard.

THREATS

A condition or object the can lead or contribute to an aircraft accident or incident.

FLIGHT TEST RISK MANAGEMENT

Flight Test Risk Management is a step in the safety assessment that aims to reduce risks to a level acceptable to the Flight Test Organization. The risk management process should be guided by the following:

- Do not accept unnecessary risks (risks that do not contribute significantly to achieve the test objectives).
- Reduce risks through actions that reduce the likelihood of hazards, minimize their effects, or reduce exposure to them.
- Conduct risk management in such a way that there is enough time to carry out risk mitigation activities before the tests are carried out (as early as possible).

Each test has peculiarities that may require the adoption of different measures to minimize risks, increase safety margins, and mitigate possible damages. Risks in flight test activities must be minimized by creating and applying a set of procedures that:

- Promote discipline and standardization in conducting the flight tests.
- Create formal processes to identify and mitigate risks.
- Contribute to increasing the safety margins of the test.

The risk management process begins when the Flight Test Plan is written, which must contain information that allows the risks of the test maneuvers to be assessed.

The risk assessment of the proposed test maneuvers must be carried out by a pilot and a flight test engineer responsible for the Test Campaign. They are the people best prepared for this assessment, due to their experience in operating the aircraft and carrying out tests. But the support of design engineering is also very important, providing relevant data and information on the aircraft's systems.

The main stages of the Flight Test Risk Management process are as follows:

- Safety Review Meeting (FTSR)
  a) Identifying hazards
  b) Estimating the effects
  c) Estimating Probabilities of Occurrence
  d) Establishment of minimum test requirements
  e) Risk minimization (calculations, simulations, preliminary tests, definition of the best sequence of activities, etc.)
  f) Mitigation actions (safety equipment, support teams, etc.)
  g) Risk classification (preliminary)
- Minimization and Mitigation actions are taken
- Procedures are incorporated into the Flight Test Orders

Those steps can be summarized in the image below.

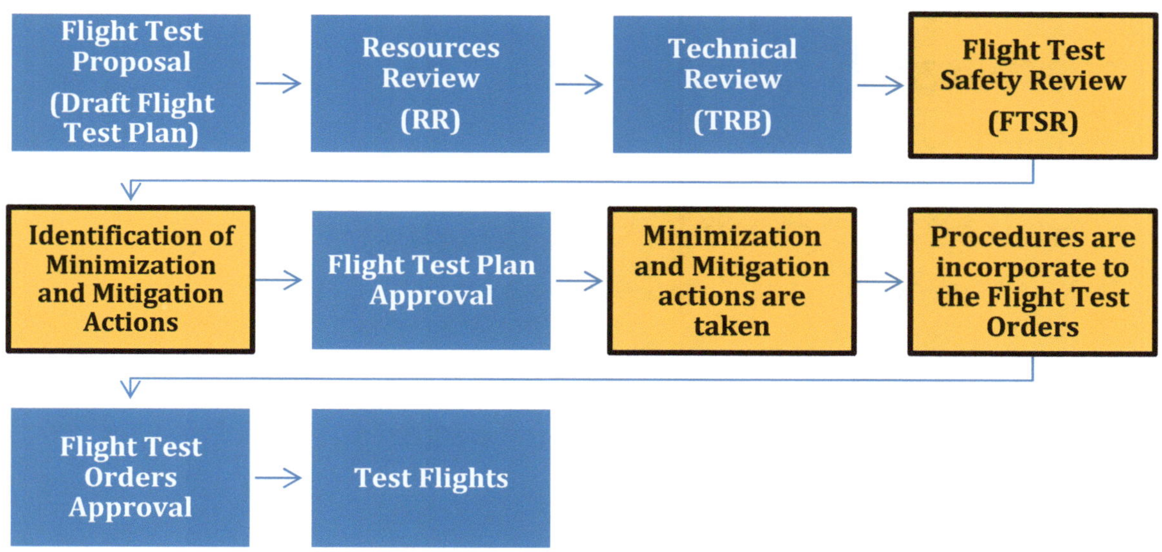

Flight Tests Risk Management (in orange)

QUALITATIVE RISK ASSESSMENT METHODS

A reliable criterion for assessing risks related to aircraft development and certification tests is needed to stimulate the process of preventive thinking and identify hazards, classify risks, and take measures to minimize them and maintain safety levels.

Government certification agencies, such as the Federal Aviation Administration (FAA) and the European Aviation Safety Agency (EASA), require aircraft manufacturers to have an internal policy to reduce exposure to risk and to allow their staff to take part in certification flights.

In addition, as of 2013, the International Civil Aviation Organization (ICAO) Annex 19 First Edition requires all aircraft manufacturers to follow a Safety Management System (SMS), which is a set of procedures to ensure that safety-related events are given appropriate emphasis throughout the organization (see Chapter 10).

Flight Test Risk Management methods are mainly concerned with risks that exist during the execution of test flights.

TEST HAZARDOUS ANALYSIS (THA)

Test Hazardous Analysis (THA) is a method widely used in Test Risk Assessment. Its basis is described in FAA Order 4040.26, which is a reference accepted by most Flight Test Organizations around the world.

THA identifies hazards associated with the execution of a test campaign. These threats are then classified, and their risk is assessed by a composite of probability of occurrence and effect of failure (see below). The risk can be:

- Low
- Medium
- High
- Unacceptable

Mitigation and minimization actions (barriers) are then proposed to reduce these risks to the lowest possible level. The probability of these threats occurring must be estimated qualitatively, if there is no statistical data, or quantitatively if statistical data is available.

Probability of Occurance						
	Highly Probable	LOW	MEDIUM	HIGH	VERY HIGH	VERY HIGH
	Probable	LOW	LOW	MEDIUM	HIGH	VERY HIGH
	Remote	LOW	LOW	LOW	MEDIUM	HIGH
	Extremely Remote	LOW	LOW	LOW	LOW	MEDIUM
	Improbable	LOW	LOW	LOW	LOW	LOW
		FAILURE EFFECT				

Determining risk using the THA method

Risk assessment must then be repeated for all the threats identified. The final risk of the test will be determined by the highest risk identified during this analysis.

This is a very practical method that usually gets satisfactory results. It also has the advantage of being well-known and accepted. However, it does not provide for debate on the testing techniques to be used.

THE "THREE FLAGS" METHOD

The Three Flags method can also be used during the development and certification of military and civil aircraft. It also applies to flight production activities.

The method is divided into three different stages that focus on the aircraft's flight envelope, the test maneuver that will be performed, and the possible failures that could affect the safety of the test. Each analysis results in a risk rating. The final risk rating is the highest of the three analyses.

Once the risk classification has been defined, a Risk Assessment is conducted to find minimization and mitigation procedures to control risk exposure. A minimization procedure tries to avoid unnecessary risks (proactive) and mitigation procedures are those that try to reduce the side effects of a known risk (reactive).

Desirable by-products of this method are the definition of the safety support for the tests, the minimum qualification of the crew, and the level of authorization to carry out the tests.

Risk assessment is carried out by analyzing the following factors:

- **Flight Envelope**
- **Test Point Execution**
- **System failures**

The Three Flags Risk Assessment methodology can be summarized in the figure below.

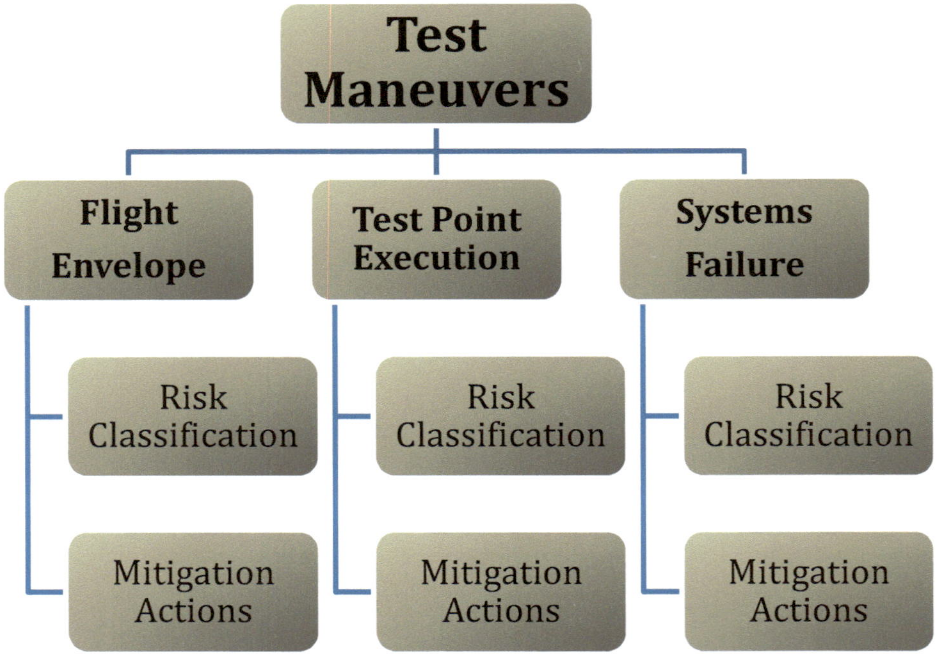

Three Flags Method

FLIGHT ENVELOPE

EXPOSURE FACTOR

The first step is to determine whether the trial should be classified as high risk, as a result of:

- Previous tests
- Available engineering models
- Dangerous or catastrophic effects that may result

The Exposure factor is defined by answering the questions in the figure below.

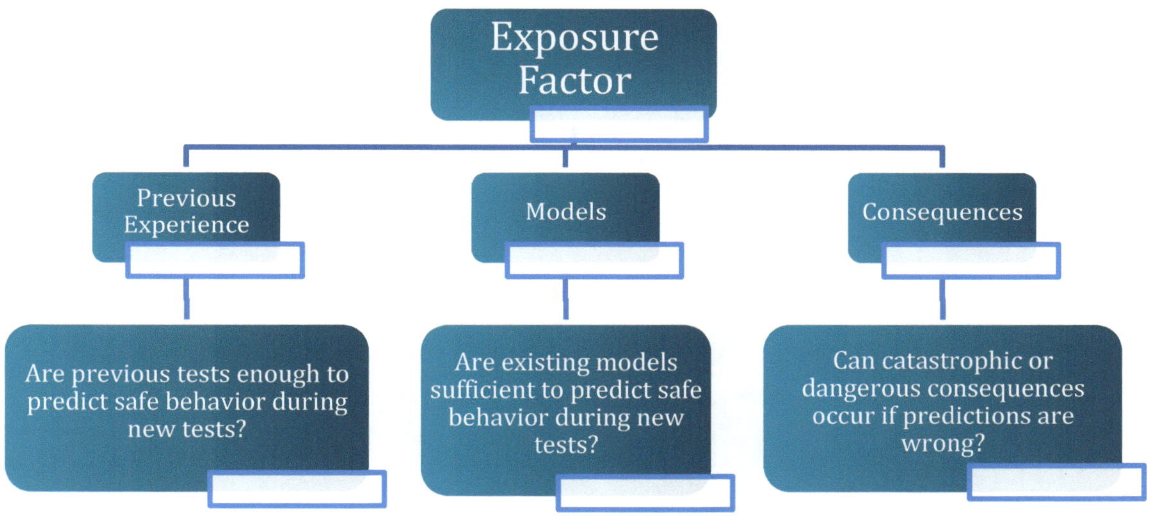

Exposure Factor

The Exposure Factor will be considered HIGH if three flags answer "yes" to the questions. If this is the case, the Flight Envelope is considered HIGH and the test will be considered high risk.

If the exposure factor is not high, the test must be classified according to one of three possible envelopes: design, operational, or limited.

- **Design Envelope**: design limits established by the aircraft's systems reports, which are normally greater than the operational flight envelope.
- **Operational Flight Envelope**: Envelope delimited by the aircraft's protections, alarms and defined in the aircraft's operational manuals.
- **Limited Flight Envelope**: Envelope that has been limited by engineering restrictions. The limited envelope occurs very often when an aircraft is under development. The

reasons that may limit the flight envelope are many and varied. They may, for example, be related to: *flutter* tests, static structural tests, flying quality tests, failures in qualification tests, etc.

The risk associated with the flight envelope will be classified according to the table below.

Flight Envelope	Risk Classification
Operational	LOW
Design	MEDIUM
Limited	MEDIUM

TEST POINT EXECUTION

The second stage of the risk assessment methodology must be carried out to determine the risk in carrying out the maneuver, as well as its recovery. It is an operational assessment.

Risk, in this case, is always related to the task to be performed, without considering the occurrence of failures.

The operational assessment will be carried out according to the following steps:

- Determine the task to be evaluated
- Evaluate the maneuver by answering the questions in the figure below.

Detecting the Need for Training or Gradual Approximation

- Training, gradual approach, or attempts are necessary **AND** can a lack of them affect safety?

Detect error tolerance

- Do test execution or positioning tolerances affect safety if extrapolated **AND** can they be considered small?

Detect the possibility of recovery or Abortives

- During the recovery or abortive maneuver, is there a likely chance of getting into an unsafe situation?

Maneuver Execution Risk

According to the answers obtained, the risk relating to the execution of the maneuver is as follows:
- If one or no flags are detected (an affirmative answer to the method's questions), the risk rating is low
- If two flags are detected (two affirmative answers to the method's questions), the risk classification is medium
- If three flags are detected (three affirmative answers to the method's questions), the risk classification is high.

The order of the flags does not alter the result, but rather the number of flags found in response to the questions.
The operational assessment should be rediscussed during the Safety Review, in which task definitions can be revised.

SYSTEM FAILURES

Failures are identified and listed based on System Safety Assessment Reports and Functional Hazardous Analysis.

However, the conditions at the time of the test and the operating situations that may affect the classification of the failure effect must also be considered.

The limitations of each prototype, documented in the Prototype Operational Limitations (POL), or the limitations of the Aircraft Flight Manual (AFM) must be considered.

The flight test engineer must compile the necessary information and carry out the system failure analysis of the relevant systems in a way that is considering the type of tests that will be carried out. The systems critical to the execution of the maneuver or systems to be tested must be analyzed.

The risk associated with system failure should be obtained by following the steps below:

A survey of failure rates relevant to the test can be found in the System Safety Assessment Report issued by Design Engineering.

The probability of failure occurring can are classified as:

- Highly Probable
- Probable
- Remote
- Extremely Remote
- Improbable.

This classification is obtained by knowing or estimating the failure rate of the system, as shown in the table below. It also depends on the category of aircraft or certification basis.

Once the failure probability is determined, the effect of the failure must be estimated to obtain the risk of system failure.

Failure Probability according to failure rate

Aircraft Category		Failure Probability				
		Highly Probable	Probable	Remote	Extremely Remote	Improbable
Aircraft Category	**Militar**	P>10-e3	10e-3>P>10-e5	10e-5>P>10-e7	10e-7>P>10-e8	P<10-e8
	FAR 23	P>10-e3	10e-3>P>10-e5	10e-5>P>10-e7	10e-7>P>10-e8	P<10-e8
	FAR 23 Commuter	P>10-e3	10e-3>P>10-e5	10e-5>P>10-e7	10e-7>P>10-e8	P<10-e8
	FAR 25	P>10-e3	10e-3>P>10-e5	10e-5>P>10-e7	10e-7>P>10-e8	P<10-e8

Failure Effect Classification: if it occurs during the execution of the test point, per the definitions set out in this Chapter (no safety effect, minor, major, dangerous, or catastrophic).

This classification, for maneuvers whose failures are not defined by the Safety Assessment Reports, must be defined in the risk assessment by the flight test engineer and test pilot responsible for it.

For failures that have already been analyzed in the System Safety Assessment Reports, the effect of the failure must be the same. A modification of the effect defined in the System Safety Assessment Report can occur if it is concluded that the effect of the failure is different in the context of the test.

MANEUVER RISK CLASSIFICATION

After being assessed by the Exposure Factor, Flight Envelope, Operational Assessment, and Systems Failure criteria, the risk of the maneuver is obtained by considering the highest rating obtained. This is not yet the final *risk*, because it must be reassessed, considering the minimization procedures that can be applied.

MINIMIZING RISKS AND MITIGATING DAMAGE

Regardless of the risk analysis method, existing risks should always be minimized to increase test safety. There are no predefined risk minimization criteria. However, some criteria are the result of experience and that allow risks to be minimized. Risk minimization procedures should be adopted considering:

- Maneuvers to be performed.
- Associated threats.
- Causes.
- Effects.

Note that procedures to minimize risk may reduce the risk level by a maximum of one level. Reductions of two or more risk levels are not recommended and must be subject to discussion and approval before they can be carried out.

Mitigation procedures must be adopted in all tests to reduce damage in the event of safety related occurrences.

Mitigation actions against damages to the crew and the aircraft should be considered, for example:
- The use of flame-retardant clothing, helmets, gloves and parachutes.
- Emergency exit device.
- Tailchute.
- Ballistic parachute.
- Minimum crew.
- Search and rescue teams, firefighters, ambulance, and medical staff on alert.

Mitigation procedures do not reduce the risk of the tests, but they do reduce the undesirable effects of existing threats.

CASE STUDIES

13

"If you want the present to be different from the past, study the past."
Baruch Espinoza

THE GULFSTREAM G650 ACCIDENT

On April 2, 2011, at around 9:34 a.m. Mountain Daylight Time, a Gulfstream GVI (G650) aircraft, FAA registration N652GD, crashed during takeoff at Roswell International Air Center Airport (ROW), Roswell, New Mexico. The aircraft was destroyed by a post-accident fire. The two flight test pilots and two flight test engineers on board were fatally injured.

Gulfstream operated the aircraft as one of five G650 flight test aircraft under an FAA experimental certificate.

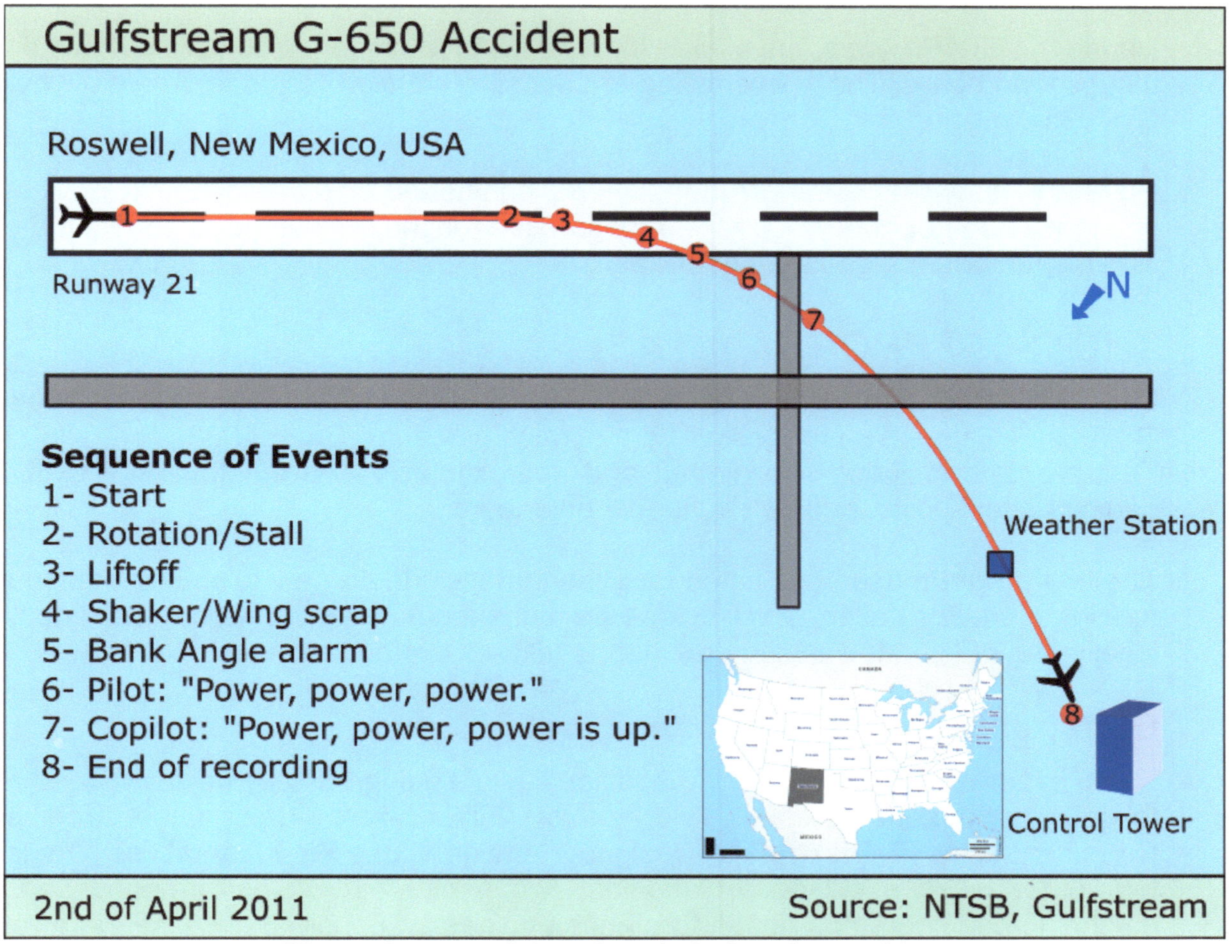

The trajectory of the aircraft during the accident

Gulfstream originated from Grumman AC, which moved from Long Island to Savanah in 1968 to build the G II aircraft.

Grumman then took over a small aircraft manufacturing company called American Aviation.

Allan Paulson bought Grumman American, which later became Gulfstream, and then Gulfstream Aerospace (1979).

Roswell was developed as a base during the GIV program due to its altitude, as there was no penalty for extrapolation up to 10,000 feet.

The Flight Test Manual was developed when Mr. Lee Johnson became the Flight Test Director for the G500 program. Until then, Gulfstream had nothing in writing (it was based on Douglas and Boeing practices). They grew out of a department that had been cut drastically by the previous president (who did not want engineering around thinking of things to test).

During the GIV Program, the aircraft had two stalls within the Ground Effect with an FAA test pilot on board. These events led to the use of *vortilons* to improve airflow over the wingtips.

However, it was a common idea among engineers that aerodynamic devices like vortilons were not good for marketing purposes (based on interviews conducted by the NTSB with former high-ranking officials).

THE G-650 PROGRAM

The G650 program began in 2005 and was officially announced in March 2008. The G650 is Gulfstream's first design from scratch since the GII in 1968. The G650 Flight Test program included five aircraft and approximately 190 employees, including 11 dedicated test pilots, and 30 Flight Test engineers.

The test program used dedicated facilities that included a new telemetry station and a mobile telemetry trailer, which allowed simultaneous missions to be supported, both locally and off-site. It was used during the take-off performance tests at Roswell.

Gulfstream G650 performance test program

Three experienced pilots (and two additional pilots being trained) were assigned for the takeoff performance tests. Two senior FTEs were Designated Engineering Representatives (DERs) or FAA Designate Airworthiness Representatives (DARs) and four other FTEs assigned. In addition, two senior aircraft performance and flight mechanic engineers and six performance engineers supported the tests on a rotating basis. The test team also included approximately 20 support personnel (including instrumentation engineers, data analysis specialists, and mechanics).

Roswell test campaign I

At the time of the accident, Gulfstream had completed all the prerequisite tests for take-off performance, including the development of a reference stall speed, determining the Ground Minimum Control Speed (VMCG) and Air Minimum Control Speed (VMCA) and Minimum Unstick Speed (Vmu).

A large part of the Minimum Un-stick Speed (VMU) testing was completed during the first series of take-off performance tests in November 2010 at Roswell, and designated Roswell I." The Roswell I flights covered 14 flights and included the first take-off performance tests of the G650.

Roswell II flight test campaign

The series of test flights designated Roswell II covered 33 flights (the last one was the crash flight) and were carried out in March and early April 2011. These tests included brake system tests, continuous take-off performance with two engines, and with the failure of one of the engines, rejected take-off, landing performance, unusable fuel and engine margin checks.

Engineering Assumptions for Take-off Performance Tests

A decrease in the IGE stall angle of 2° was used in previous Gulfstream models. Historically, this assumption provided adequate safety margins, since tail power limitations forced take-off speeds to be higher. It was believed that this reduction would also approximate the G650's actual reduction. For the G650 test program, it was assumed that the maximum lift coefficient (CLmax) in ground effect was equal to the CLmax in free air.

The Crew

The 64-year-old pilot-in-command graduated from the United States Military Academy, received an MS from the Air Force Institute of Technology, and held an airline transport pilot (ATP) certificate for landplanes, gliders, single-engine and multi-engine aircraft with type ratings for the Gulfstream GIV, Gulfstream GV, Gulfstream G-1159, Boeing B-707, and Boeing B-7204. According to company records, the pilot-in-command had accumulated 11,237 total flight hours, including 9,500 hours in command. He accumulated 263 hours on the Gulfstream GVI (of which around 160 hours were in command). The PIC was a distinguished graduate of the United States Air Force Test Pilot School. He was a test pilot and Designated Engineering Representative (DER) for the FAA.

The 51-year-old co-pilot graduated from the United States Coast Guard Academy and held an ATP certificate for single-engine and multi-engine landplanes with type ratings for the Gulfstream GV, Gulfstream G-100, Boeing B-737, and Westwind Astra IA-1125. According to company records, the co-pilot had accumulated 3,940 total flight hours, including 2,637 hours in command and 140 hours on the Gulfstream GVI (78 of which were in command). He also held flight instructor certificates for gliders, single-engine landplanes, and multi-engine landplanes, and commercial pilot certificates for single-engine landplanes and multi-engine airplanes. The co-pilot graduated from the United States Naval Test Pilot School (USNTPS). Although he was relatively new to Gulfstream's large-cabin aircraft, within the company he was considered highly qualified, positive, and appropriately assertive.

A 48-year-old Test Engineer who earned a Bachelor of Science degree in Aerospace Engineering from Iowa State University and had approximately 20 years of Flight Test experience. His experience included stall speed and take-off performance tests on the MD-11 program, stall speed and climb performance tests on the C-130J program, and high AOA and maximum take-off weight tests on the F-16 program. He was head of the Flight Sciences section at another aircraft manufacturer from 1999 to 2002, overseeing aircraft performance tasks. He was an FAA-accredited representative (DER).

The second flight test engineer was 47 years old and had been with the company since 2006, where he started as a production test pilot for the G150 and G200. The day of the accident was his first time participating in the field performance campaign. He had been responsible for the G-650 airspeed calibration tests. Prior to Gulfstream, he worked at Lockheed Martin on the C-130J and Galaxy C-5M programs. Also as first officer and then captain on CRJ-200s for Atlantic Southeast Airlines. He also worked as a contract test engineer on Gulfstream's GV program and as a project engineer for the FAA. At Douglas Aircraft, he worked on the T-45A and MD-11 projects in the performance and flight qualities department. He also served in the United States Army Reserve as a CH-47D and UH-1H pilot.

The Prototype Aircraft

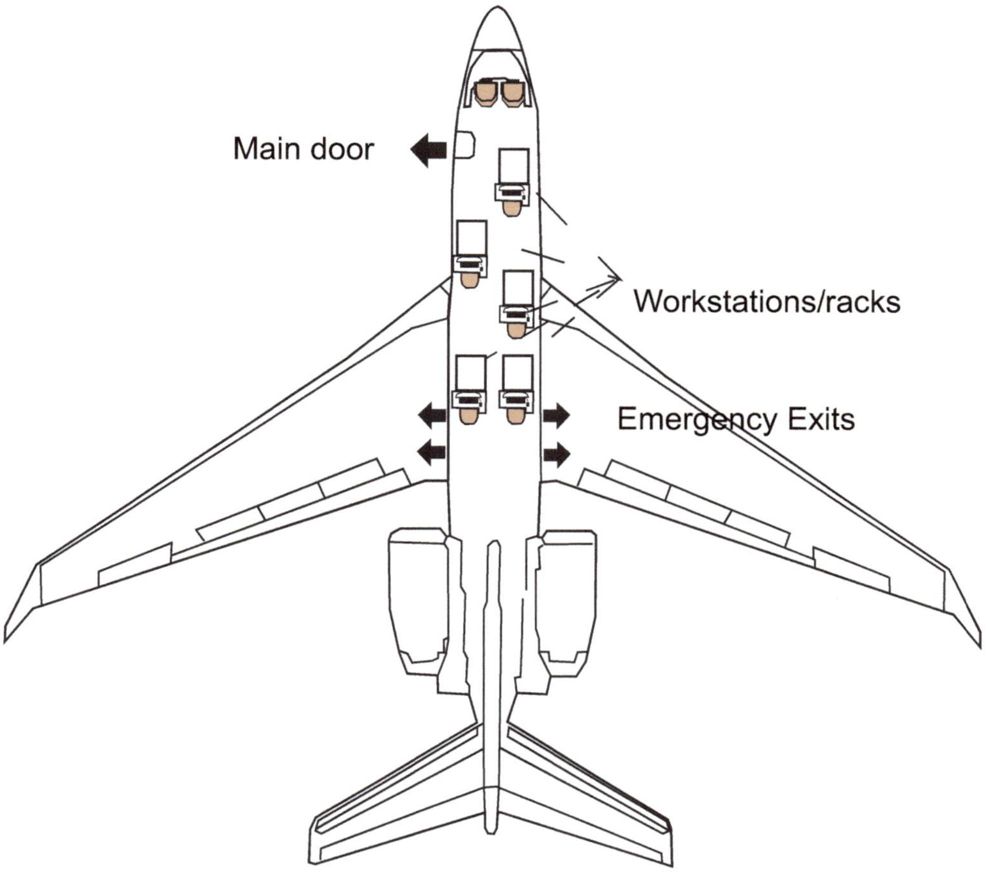

Diagram of the Cabin Interior of the Crashed Aircraft (6002)[24]

[24] Source: NTSB/Gulfstream AC. Not to scale.

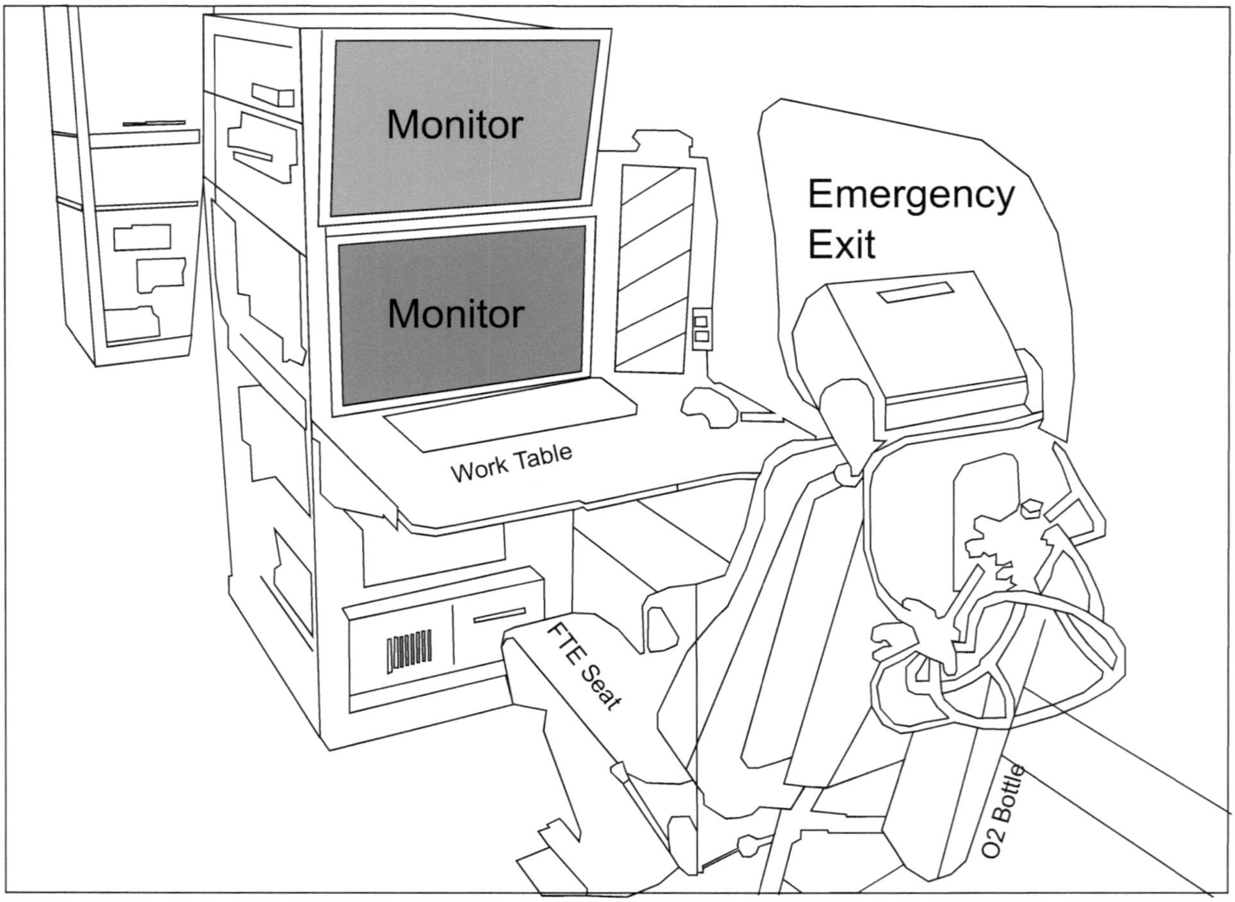

Flight Test Engineer's workstation on board the test aircraft

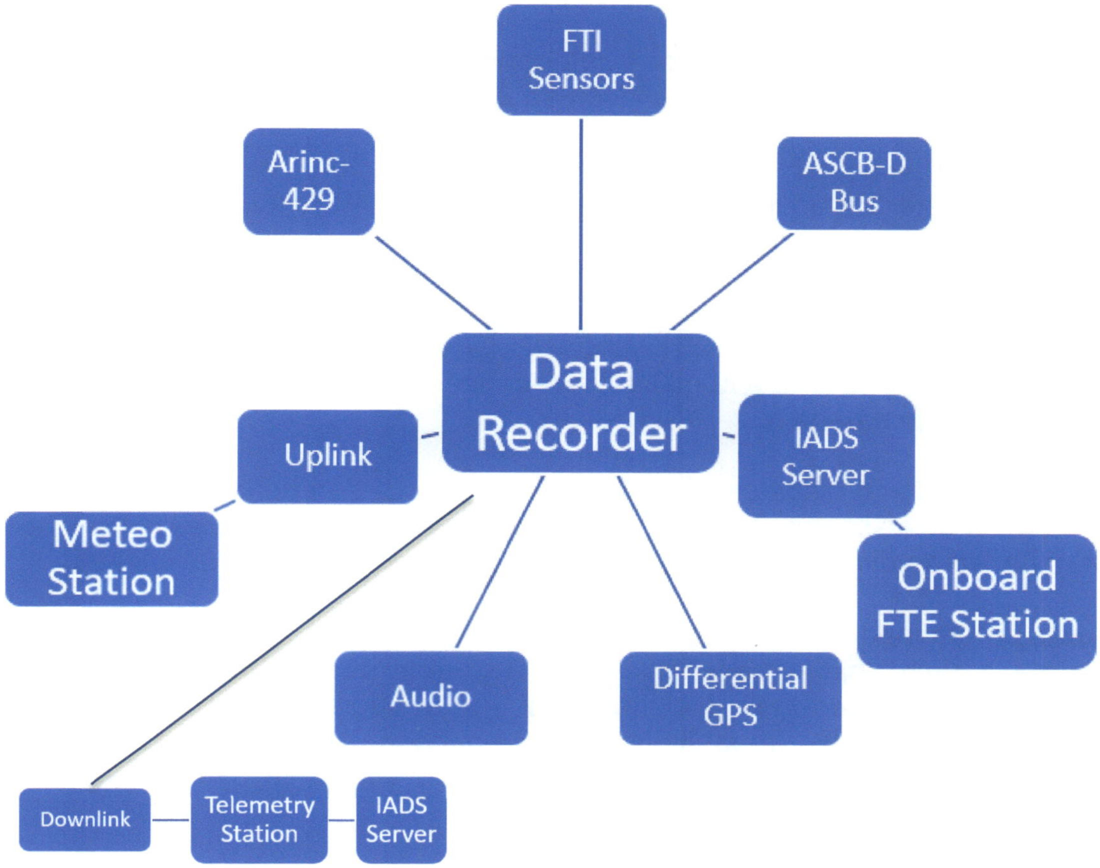

G-650 Data Recording System

EMERGENCY SUPPORT

The firefighter team had an Index A category and Roswell Airport maintained two ARFF vehicles at the airfield, Crash 1 and Crash 2.

Crash 1 had a capacity of 1,000 pounds of Purple K dry chemical powder and 1,500 gallons of water for aqueous film-forming foam (AFFF) production supplied by a 200-gallon foam concentrate tank. The unit was equipped with a roof tower and dual-agent hand hoses.

Crash 2 had a capacity of 500 pounds of Halon and 1,500 gallons of water for AFFF production supplied by a 200-gallon foam concentrate tank. The unit was also equipped with a dual-agent hand hose.

According to local dispatch records and the Fire Department's incident report, ARFF responded to the scene of the accident within four minutes. All times in the dispatch logs were based on dispatcher entries manually entered the system.

It was determined that Roswell Airport and its firefighting team were not informed that the tests were high risk. The firefighting team's readiness status would have considerably reduced their response time. The accident report emphasizes that in this accident, this would probably not have changed the fatalities, but that in other circumstances, the response time of firefighters can have a major influence in the consequences of an accident.

PERFORMANCE ANALYSIS

A simplified analysis of what happened from a performance point of view can be discussed. One important fact is that there is a significant reduction in induced drag when within the ground effect, where the height above the ground is less than an aircraft's wingspan. This effect has an exponential decay until it is practically nil when a height equivalent to a wingspan is reached.

There are three important consequences of reducing the drag induced by the ground effect:
- Increase in the coefficient of lift for a given angle of attack
- Reduced stall angle of attack
- Reduction of the required thrust

Estimates made using CFD by Gulfstream show, because of the ground effect, a reduction in the angle of attack at stall of up to 3 degrees (see Figure below).

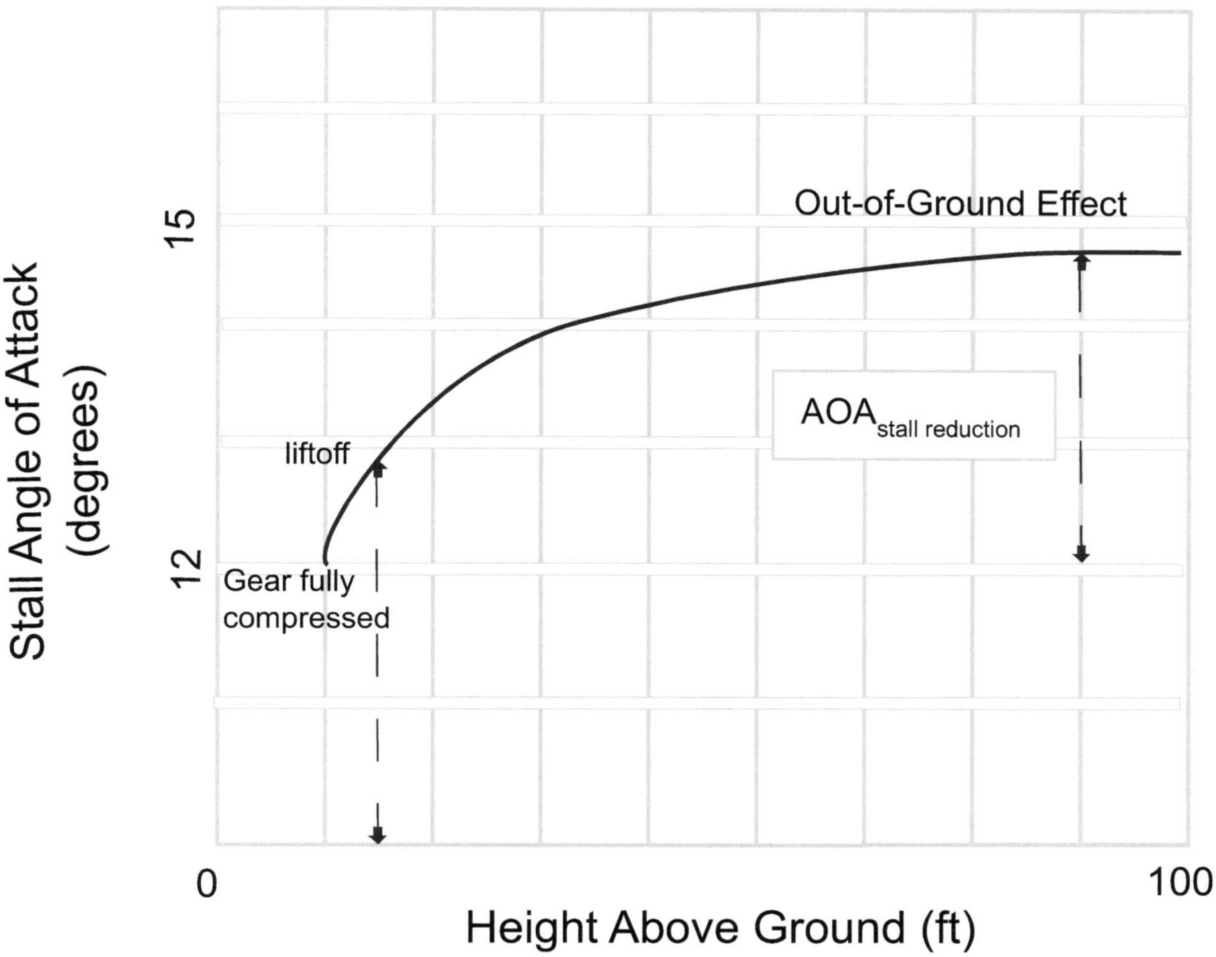

AOA of Stall as a function of height above ground (source: Ref.02)

The aircraft's stall warning system was studied using the data recorded during the test. They show that the stick shaker was activated at the angle of attack predicted for stall in the ground effect, therefore not having any anticipation to avoid stall.

The "normal" method uses Vmu as a "base" and then "works forward" to determine the operating speeds. The speeds are then increased by considering the VR/Vsr, VLOF/Vsr, VLOF/Vsr, and V2/Vsr ratios (figure below). V2 shall also consider the minimums in the certification regulations (V2 must be equal to or greater than 1.13 VSR).

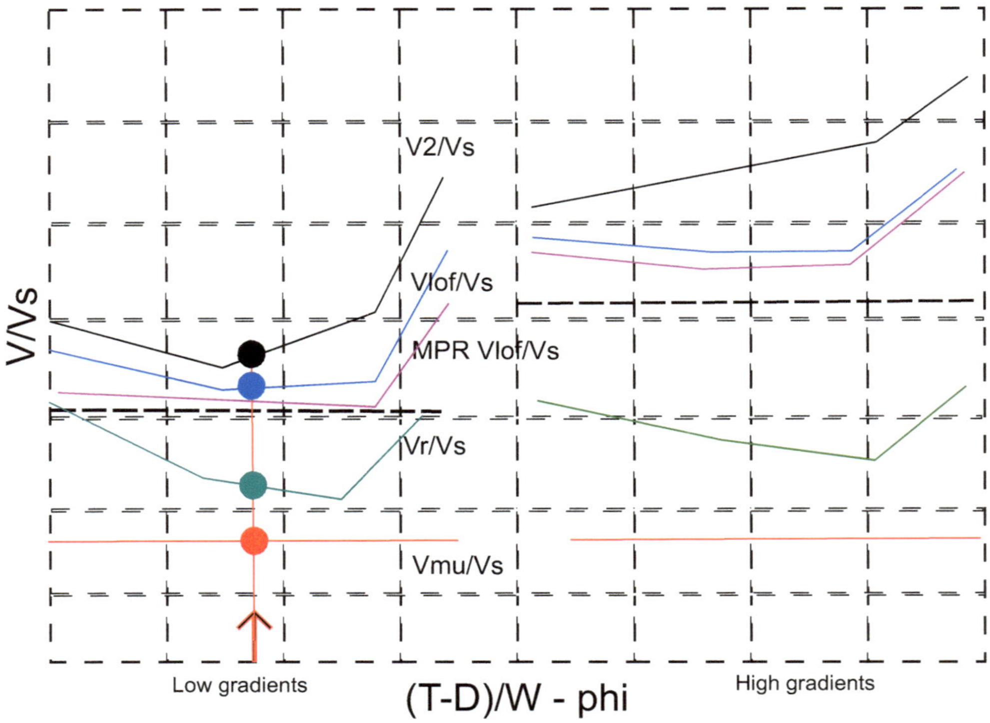

Commonly used method of setting take-off speeds.

In the G650 program, V2 was established as the minimum required by regulation (1.13 VS), and decrements were then subtracted from V2 "backward" up to VLO and VR. This method led to VR speeds below the requirement of VRmin ≥ 1.1 VS.

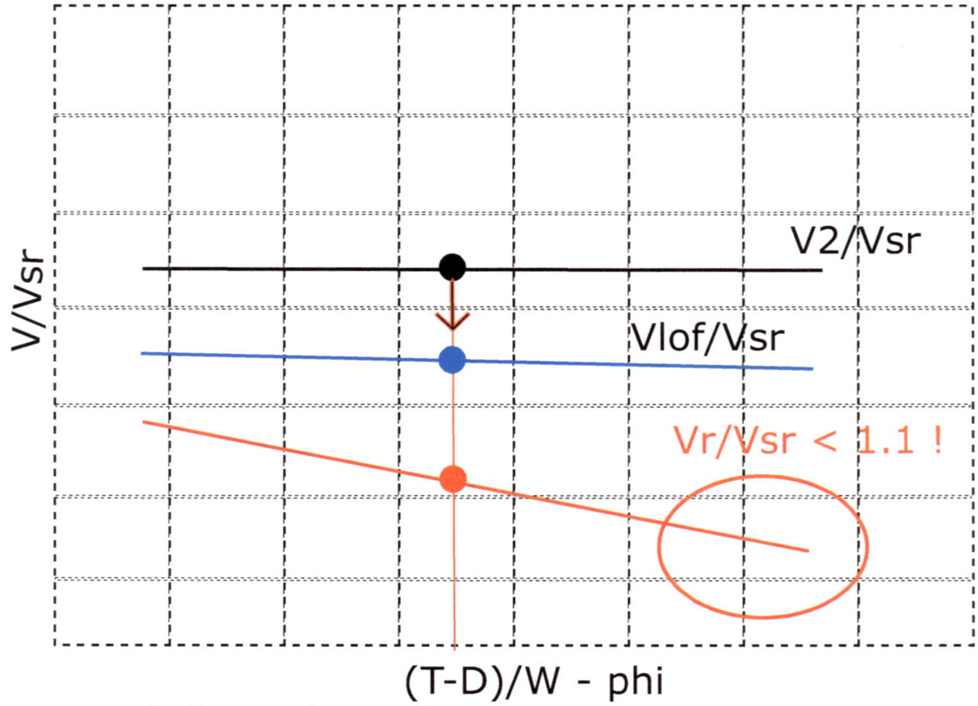

Definition of take-off speeds in the G650 Program

CONCLUSIONS

The NTSB concluded that the probable cause of the accident was a stall followed by uncontrolled roll during a single-engine takeoff test. The main causes of the accident were:

1) Gulfstream's failure to correctly develop takeoff speeds and correct takeoff safety speeds (V2) during pre-accident testing.

2) The persistence and aggressiveness of the flight test team in achieving takeoff speeds that were incorrect and too low.

3) Gulfstream's failure to investigate previous uncommanded roll events that indicated that the estimated in-ground effect stall angle of attack was too high.

Gulfstream's failure to manage the program with an aggressive schedule without ensuring that roles and responsibilities were well defined, engineering processes had sufficient technical planning and oversight, and threats were properly identified and controlled was considered a contributing factor to the accident.

THE GROB G180 ACCIDENT

BACKGROUND

Grob Aircraft is a German company specialized in the production of gliders and general aviation aircraft. The company has a great deal of experience in the production of composite materials. Grob Aircraft originated from Grob Aerospace, founded in 1971 by Burkhart Grob. It initially grew out of its founder's vision of producing gliders at a reduced cost and production time by the standards of the time. He managed to secure a contract to produce Cirrus gliders under license and later, in 1974, began producing his own gliders. The first of these was the Grob G-102 Astir, followed by the G-103 Twin Astir for two crew members.

At the end of the 1970s, Grob began designing and developing motorized gliders. The G-109 would be the first motor glider made entirely of composite and received FAA certification in 1981. It was used by the Royal Air Force until 2018 to train cadets.

After the G-109, Grob produced the G-115, which received FAA certification in 1987 and the GF-200 and the G-160 Ranger, which did not go into production. In the 1980s, Grob partnered with the US company E-Systems and Garret Engines to produce the G-520, which broke five world altitude records.

The Grob G180 project began after 2000 at the company's headquarters in Tussenhausen-Mattsies, Germany, in response to market demand for small business jets. It was a twin-engine jet aircraft with seating for up to eight passengers and made with extensive use of composite materials. Development was suspended in 2008 following the company's insolvency.

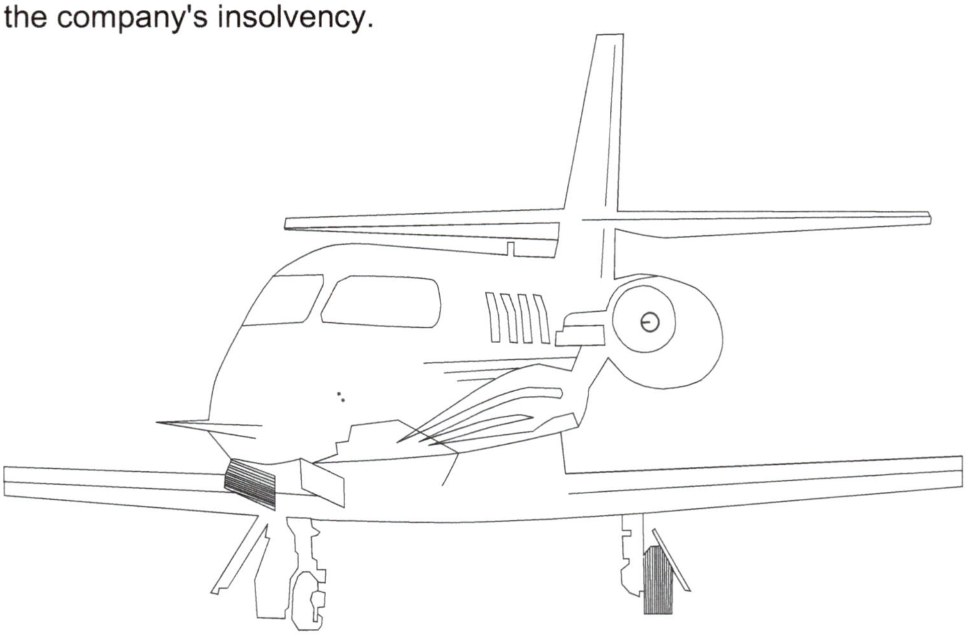

Grob 180 Aircraft

DESIGN SPEEDS

The Grob G180 was designed for a cruising speed of 270 knots and Mach 0.70. Maximum altitude was 41,000ft. Due to certification requirements and design margins, the aircraft had to demonstrate in flight that it was *flutter-free* up to the design speed (Vd) of 338 knots. It was also required to demonstrate a 20% margin over Vd (406 knots) through analysis.

FIRST PROTOTYPE (P1)

The first flight took place in 2005. The flight envelope was gradually expanded. It reached a speed of 280 knots based on GVT data after about a year. However, it was found to have limited elevator authority, which required a design change.

SECOND PROTOTYPE (P2)

The second prototype was an aircraft designed to develop avionics systems and envelope expansion tests were not planned for it. It had a horizontal stabilizer 16% larger than the P1.

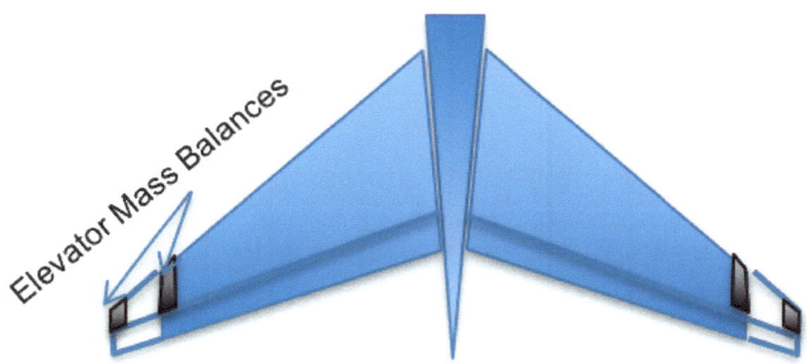

Horizontal stabilizer and balancing masses

In November 2006, the second prototype suffered a fatal accident caused by flutter on the horizontal stabilizer during a demonstration flight. The root cause of the accident was not determined but the investigators believe the elevator mass balance was probably the cause of the accident. The chief test pilot of the company was killed in the accident.

The prototype number two had an airspeed limitation of 200 kt, which was lower than the 270 kt of prototype number one. However, during the flight display airspeeds up to 270 kt were flown (based on radar information).

No flight test instrumentation or Flight Data Recorder was recorded during the flight which made it very difficult to understand the sequence of events. Several efforts were made to retrieve data from the aircraft fadecs and avionics but without much success.

The flutter characteristics of the second prototype were estimated by Ground Vibration Tests (GVT) analyzed by a contractor company. The initial analysis has shown that the flutter speeds were too low, and the elevator mass balance was increased from 1 kg to 4.5 kg.

The first prototype horizontal stabilizer had shorter span than the second prototype and had already been flown to Vd and logged more than 200 flight hours.

The weather conditions at the time of the accident were not according to the company policy which required at least 1500 ft cloud ceiling. The conditions during the demonstration were 1000 ft cloud ceiling.

It is estimated that the horizontal stabilizer failure occurred in less than 1 second. The failure of both horizontal stabilizers was similar which indicates a symmetrical flutter mode.

LESSONS LEARNED

- Configuration Control is a critical safety concern.
- Limitations can be very different between two prototypes and must be controlled very closely. Each prototype can be unique regarding handling characteristics, performance and flutter.
- Demonstration flights must go through the same safety process as flight tests when a prototype aircraft is used.
- Flight Test Instrumentation (FTI) is crucial during the development phase of a new airplane. However, the FTI is not crashworthy and therefore, a telemetry system is normally required to have a backup in case of accidents.
- Aircraft systems that have recording capabilities are not crashworthy and are not appropriate to support accident investigations.
- The lack of data to support conclusions related to safety events has a huge effect in the time response of the company to overcome the issue e resume flying. This must be considered when decisions to reduce or remove the FTI are made.
- A single point failure of the flight controls mass balances can cause a catastrophic failure.

THE SINO SWEARINGEN ACCIDENT SJ30-2 002

On April 26, 2003, at 10:05 a.m. Central Daylight Time, a Sino Swearingen Aircraft Corporation (SSAC) SJ30-2, N138BF, serial number 002, was destroyed when it impacted terrain near Loma Alta, Texas.

The test pilot was fatally injured. Visual meteorological conditions prevailed for the flight, which departed from San Antonio International Airport (SAT), San Antonio, Texas.

THE AIRCRAFT

The SJ-30 concept was made with advanced technology and relatively high performance but was a low-cost entry-level corporate jet. It would compete in the market against the Cessna CitationJet/CJ1 and the Raytheon Premier I.

The first prototype of the SJ-30 flew for the first time on February 13, 1991, and development progressed slowly until the announcement in 1994 that Swearingen would form a joint venture with Taiwanese investors to create Sino Swearingen to build the SJ-30 in a new factory in Martinsburg, West Virginia.

Sino Swearingen decided to extend the basic SJ-30, resulting in the stretched SJ-30-2 with an increased wingspan of 1.32 m (4 ft 4 in). The prototype SJ-30 was modified to represent the SJ-30-2 and flew in this configuration in November 1996. The definitive and more powerful FJ44-2A turbofans were installed on this aircraft in September 1997. Three new production-representative prototypes were scheduled to carry out a 1,400-hour flight test program (the first of which flew on November 30, 2000), paving the way for certification and the first deliveries in the third quarter of 2003.

The SJ-30-2's features included a Honeywell Primus Epic avionics suite with three 20 x 25 cm (8 x 10 in) color LCDs and integrated IC-615 avionics computer.

The engines were two 10.2kN (2300lb) Williams FJ44-2A turbofans.

The aircraft had a maximum take-off weight of 5,987kg (13,200lb).

It was designed for a maximum operational mach number (Mmo) at Mach 0.83 and maximum altitude of 49,000 feet.

THE TEST CAMPAIGN

The flutter test was to be conducted in two phases, with the first phase planned to release the airplane to its "maximum operating speed" (Vmo/Mmo) of 320 KCAS/Mach 0.83, and the second phase, to release it to its Vdf/Mdf of 372 KCAS/Mach 0.90.

Phase 1 of the flutter tests to Mmo (Mach 0.83) had been completed. And the first flutter mission of phase 2 was carried out the day before the accident, with the same pilot on board. This flight aimed to complete flutter test points at Mach 0.844 and Mach 0.864.

However, the first flutter mission was terminated after an uncommanded roll to the left.

The pilot also reported that during that flight he had felt a "rumble" in conjunction with the left roll at, according to his notes, indicated Mach 0.855.

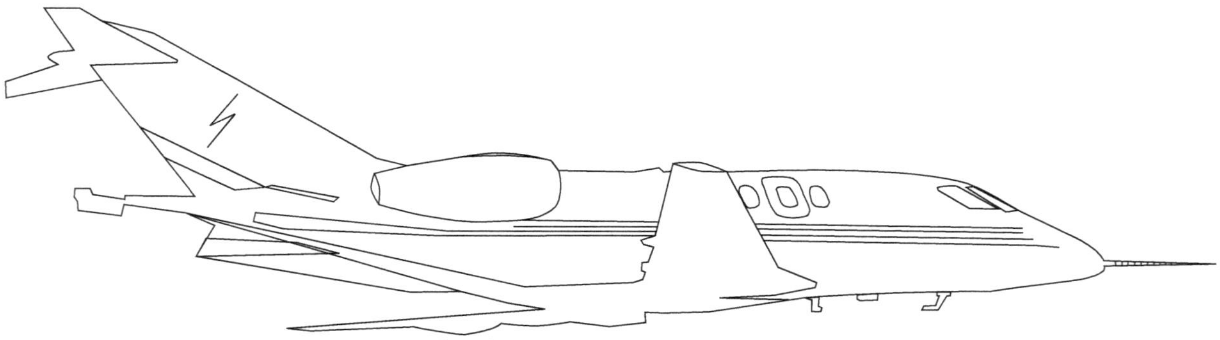

SJ-30-2 prototype aircraft

A discrepancy was also noticed between the speeds displayed and those reported by the chase pilot. But after the flight, the pilot realized that he had incorrectly set up the airspeed display.

Before the accident, the company estimated the high-speed stability and control characteristics of the SJ30-2 by:
- extrapolating low-speed wind tunnel data,
- using methods from the USAF Stability and Control Data Compendium (DATCOM),
- conducting numerical simulation with Computational Fluid Dynamics (CFD) tools and
- extrapolating flight test data.

THE TEST FLIGHT

Prior to the test flight, a mission briefing was conducted via teleconference between San Antonio personnel and telemetry van personnel.

The flutter consultant advised the pilot that if he ran out of aileron/elevator trim, the tests could still be completed, even if the pilot had to hold aileron/elevator force to control the airplane. But he also stated that it was pilot's decision as to continue or not.

After take-off, the airplane climbed to 39,000 feet and set up for a dive targeting indicated Mach 0.884.

The pilot then initiated elevator, aileron and rudder impulses (also called jerks). Telemetry van personnel noticed that all the modes excited were "well damped.."

After the pulses were completed, the pilot stated that the uncommanded roll to the left (which was experienced on the previous flight), did not occur. It was reported that prior to reaching the test point of indicated Mach 0.884, the pilot had to use full trim and could not let go on the wheel.

After climb back up to 30,000 ft and with telemetry lock, the test pilot began a shallow dive again.

At indicated Mach 0.884, each control surface was again pulsed by the pilot, and the responses were again "well damped.."

Following the final pulse, the pilot was cleared to the next test point, however, the pilot did not acknowledge the clearance but instead reported that the airplane was rolling to the right, and he could not stop it.

After the test airplane began to roll, and the test pilot stated that he could not stop it, the chase airplane pilot called him to get out twice. The test pilot responded that he could not get out, that there were too many "g's."

After a few seconds, the test airplane entered a "barrel-roll type maneuver" to the right, then continued to roll and increased its dive angle until ground impact.

The accident pilot did not say anything further about how the airplane was performing, or what he was experiencing.

Data recorders on board the airplane were recovered; however, their condition precluded any data recovery. The telemetry data included only 27 parameters, most of which were accelerations.

FACTS OBSERVED DURING THE INVESTIGATION

The aircraft lateral control was the focus of the NTSB investigation. The aircraft's roll trim system used an adjustable trim spring to apply a constant force to the yoke. "Because of this characteristic the amount of trim required to balance an aerodynamic force asymmetry depended on speed," the Safety Board said.

In May 2002, the company issued a temporary limitation restricting the use of the aileron trim by the pilot in the range of 20 to 80% because the aileron trim power unit stalled at approximately 14% and 92% of the travel.

A speed restriction of 250 knots was imposed in June 2002. The ailerons were removed, measured and replaced, but no issues were found.

Sino Swearingen later concluded that the plane's tendency to roll to the right could be attributed to torsional deviations of the wing and the other aileron in the opposite direction to the design.

In October 2002, the speed restriction was changed to 320 knots, or Mach 0.83, following the completion of the Phase 1 flutter test. The tendency to roll to the right was accepted even though almost full roll trim was required at that airspeed.

The use of tuffs on the upper surfaces of the wings confirmed the presence of "large regions of separation induced by shock waves above Mach 0.81." In April 2003, the airbrake travel was limited from 35 degrees to 17.5 degrees to avoid undesirable rolling characteristics.

To improve lateral control the engineers fitted the tip of the accident airplane's right wing with a Gurney flap, an aerodynamic device designed to balance the airplane on the lateral axis, regardless of speed.

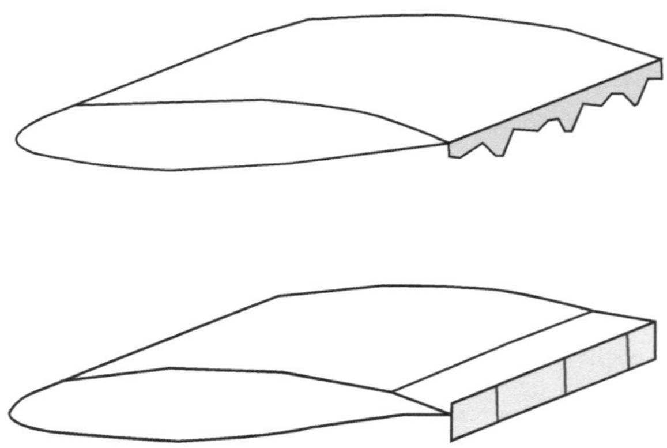

Examples of Gurney Flaps on airfoils

The Gurney flap improved the lateral trim margin and, for airspeeds of up to 305 kcas, approximately 40% lateral trim was required to balance the airplane.

Despite the effectiveness of the Gurney flap, Sino Sweringen engineers considered that the airplane would still require left wing downward control to compensate laterally as airspeed increased beyond Vmo (320 knots). The decision was to continue with the flutter test if the pilot needed a small amount of lateral force above 320 knots.

Rudder pedal was also used to improve lateral control between Mach 0.84 and 0.86 when the aircraft descended from 34,000 to 30,000 ft.

During one test flight the aircraft went into an uncommanded spin (left wing down). The rotation was corrected by the pilot's stick for about 20 seconds while the airplane decelerated below Mach 0.85. Rudder was also used by the pilot to regain lateral control during the recovery.

Low-speed wind tunnel tests, carried out before the accident, revealed that the aerodynamic brake at high angles of attack (post-stall) would reduce the lateral stability of the wing.

During the accident flight, as the accident aircraft accelerated towards the target Mach number, it transitioned from level flight to a trajectory angle of around -7 dg and then about -10° at the conclusion of the aileron pulse. The trajectory angle continued to decrease to an estimated final value of -77 degrees at the impact with the ground.

AFTER THE ACCIDENT

The company developed a test plan to define the high-speed stability and control characteristics of the SJ-30-2.

Wind tunnel data provided evidence that roll authority deteriorated above Mach 0.86. As a result of the wind tunnel tests and other post-accident research, Sino Swearingen added vortex generators to the wings to delay the onset of shockwave-induced separation and installed thicker trailing edges on the ailerons to improve the effectiveness of the ailerons at high Mach numbers. In addition, a high Mach number spoiler system would increase lateral control.

Flight tests of the SJ-30-2 resumed three months after the accident, and the modifications mentioned were incorporated into the S/N 004, which passed all the high-speed tests in August 2003. S/N 004 was fitted with a high-speed drag parachute.

Although the engineering analysis predicted positive lateral stability up to Mach 0.90, lateral control was lost during the accident flight, and the airplane rolled about seven times over 49 seconds, from about 30,500 feet to an almost vertical impact on the ground.

A 1/9 scale model was tested in a high-speed wind tunnel. The design of the model allowed measurements of the articulation moments of the horizontal stabilizer, aileron, elevator, rudder, external spoilers, and aerodynamic brake. In addition, vortex generators, ailerons with thicker trailing edges, Gurney flaps, winglets, and strakes were also tested. In January 2004, the tests were carried out in a 2.5 x 2.7 m transonic tunnel in Bedford, England.

The wind tunnel test data indicated that the lateral stability of the SJ-30-2 deteriorated with increasing Mach number and angle of attack. Lateral stability, measured in terms of rolling moment coefficient due to sideslip (called $Cl\beta$), became positive (unstable) above Mach 0.83. Because of this, a rudder input aiming to lift a low wing could instead, beyond a certain Mach number, made the situation worse. In the same way, an elevator input to recover from dive would tend to increase the angle of attack, also resulting in a deterioration in lateral stability.

The flow visualization results revealed that the upper wing surface flow separated between Mach 0.84 and 0.88, and the lower wing surface flow separated between Mach 0.86 and 0.88, at an angle of attack of 2 degrees and a slip angle of 0 degrees, which were representative of the flight condition during the accident.

Computational Fluid Dynamics (CFD) methods were used for the wing design and to complement the SJ-30-2's high-speed stability and control database. Before the accident, Vortex Lattice (VLM) and Euler methods were mainly used. Euler methods tend to predict shock wave's locations further aft than actual locations during transonic flight conditions.

The wing design calculations for the SJ-30-2 were carried out using WIBCO, a NASA/Grumman small perturbation transonic code. But the code lacked an asymmetric analysis capability. It was mainly used by Sino Sweringen for cruise analysis to validate the onset of separation.

Two-dimensional CFD studies of the aileron indicated that aileron power would decrease with increasing Mach number.

After the accident, SSAC performed inviscid calculations up to Mach 0.9, including lateral control, to understand three-dimensional, transonic, and asymmetric characteristics. A more advanced, fully viscous Navier-Stokes CFD code was also used to obtain additional information and to improve stability prediction and control derivatives.

The results of the high-speed flight tests on S/N 004, which incorporated the configuration modifications described above, demonstrated improved high-speed stability and control. The airplane flew several test points to Vd/Md (372 KCAS/0.90 Mach). The modified SJ-30-2 configuration demonstrated positive lateral stability at Mmo (0.83 Mach) and neutral lateral stability at approximately 0.85 Mach.

Recoveries from the high-speed dive, achieved by reducing the engine to idle, resulted in a return to a laterally stable flight regime in a few seconds. Releasing rudder control from a stabilized skid condition returned the airplane to level flight at all Mach numbers tested up to Mach 0.90, even when the roll coefficient due to sideslip was positive (unstable).

Sino Sweringen successfully completed high-speed and flutter tests of the SJ-30-2 in August 2004 and demonstrated that the roll *spoiler* at high Mach numbers was not necessary.

CONCLUSIONS

1) Predictions of high-speed flight characteristics are also mandatory for high-speed flutter tests, in addition to flutter vibrations.

2) Stop the high-speed/flutter test immediately if control instabilities or loss of control tendencies are observed.

3) Roll tendencies that increase with increasing speed or Mach are strong indications of serious aerodynamic problems and can lead to loss of control at high speed. The loss of control is aggravated because when the roll begins, the aircraft tends to accelerate even more.

4) Check how the high-speed flight qualities were projected (wind tunnel, CFD, etc.).

5) Unstable lateral-directional characteristics above a certain Mach may contraindicate the use of the rudder to correct rolling tendencies ($Cl\beta > 0$, unstable).

6) Characteristics that tend to decrease lateral stability as AOA increases can make recovery very difficult.

7) Unsatisfactory flight characteristics may indicate the need to refine aerodynamic models or use a transonic wind tunnel. In addition, they may indicate the need for a high-speed tailchute.

THE SPACE SHUTTLE COLUMBIA ACCIDENT

On February 3, 2003, during re-entry into the Earth's atmosphere, the Space Shuttle Columbia disintegrated above Texas. The *Space Shuttle* was destroyed, and seven astronauts lost their lives.

The shuttle had completed a 16-day mission when the accident occurred.
The physical cause of the accident was determined to be damage to the left wingtip due to foam coming loose from the external fuel tank (ET) during take-off.

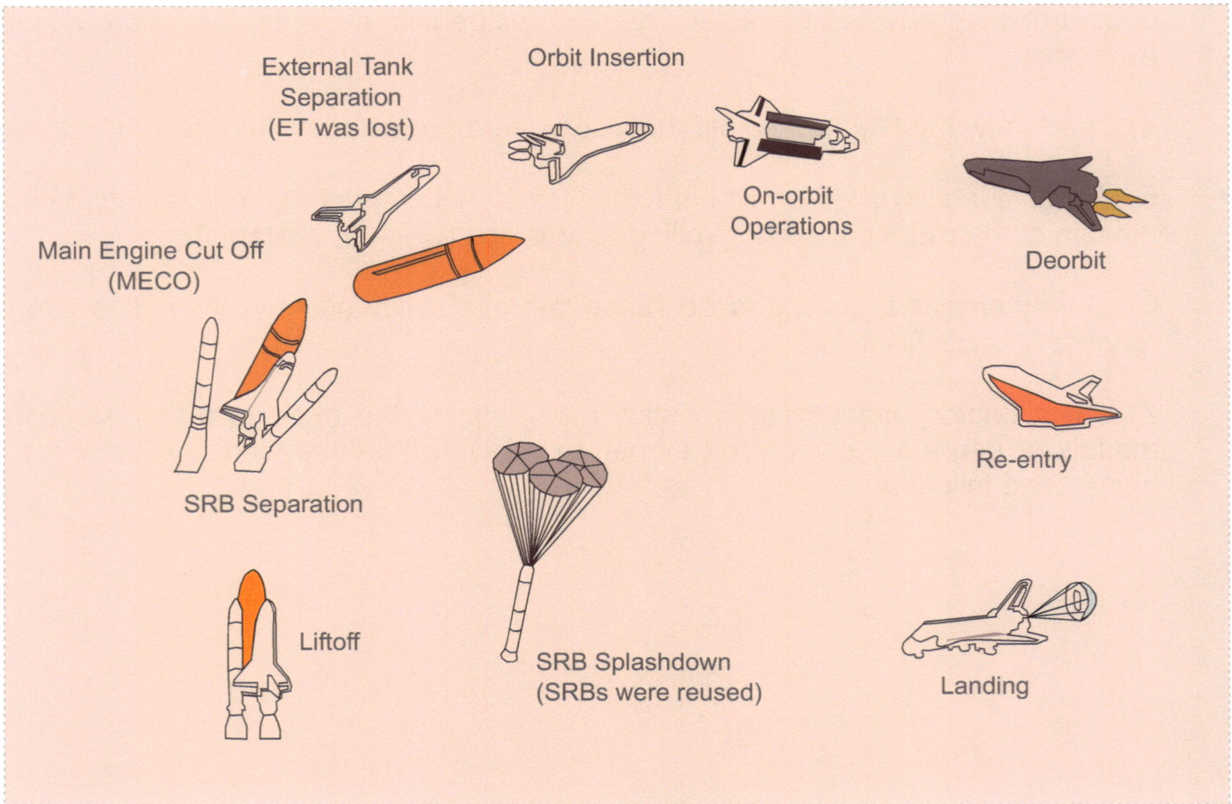

Typical Space Shuttle mission[25]

NASA

"To provide research into problems of flight within and outside the Earth's atmosphere, and for other purposes."

With this simple preamble, Congress and the President of the United States created the National Aeronautics and Space Administration (NASA) on October 1, 1958.

It was a response to the launch of Sputnik. Its main projects were:

[25] Source: NASA Columbia Accident Report Volume 1

- Project Mercury
- Project Gemini
- Project Apollo
- Space Shuttle Project
- International Space Station

In addition to the space flight program, NASA also maintains an active (albeit small) aeronautics research program, a space science program (including deep space and interplanetary exploration), and an Earth observation program.

The agency also conducts basic research activities in various areas. NASA has several Research Centers where its programs are carried out.

The centers provide the buildings, facilities, and support services for the various programs. The programs, along with the research centers and headquarters, hire public employees and private sector contractors to support their undertakings.

THE SPACE SHUTTLE PROGRAM

Discussions about the Space Shuttle program began to define what would follow Project Apollo. The agency hoped to build a space station for 12 people by 1975; subsequent stations would hold 50, then 100 people.

At the same time, NASA would develop the capacity for manned exploration of Mars. NASA wanted to develop a fully reusable vehicle.
NASA's vision of a constellation of space stations and trips to Mars had little connection with the political realities of the time (Presidents Johnson and Nixon administrations).

With NASA's Space Station plans postponed and production of the Saturn V rocket canceled, the Space Shuttle became the only manned space flight program the agency could carry out. But a new justification for the Space Shuttle was needed. To overcome the objections of the White House Office of Management and Budget, NASA began to justify space exploration on economic grounds.

By combining commercial, scientific, and national (Department of Defense) payloads, 50 Space Shuttle missions a year would be needed. This was enough to justify the investment in the Space Shuttle Program. This is approximately one mission per week.

Meeting the military's perceived needs and, at the same time, keeping mission costs low posed huge technological obstacles. These missions and cost requirements called for a revolution in space technology.

The space shuttles would be the first reusable spacecraft, the first to have wings, and the first to have a reusable thermal protection system. They would be the first to fly with reusable engines using high-pressure H_2O_2, and the first winged vehicles to transition from orbital speeds to a hypersonic glide during re-entry.

In 1971, NASA was informed that it could count on a maximum of $5 billion over five years for any new program.
In 1972, NASA told the White House that for $5.15 billion ($37.8 billion in 2024) it could build a vehicle that could meet all the performance requirements with a lifespan of 100 missions per vehicle, at a cost of $7.7 million per flight ($51.3 million in 2024).

The Space Program's suppliers

The Space Shuttle Program employed many suppliers throughout the USA. The main ones were:

United Space Alliance
An alliance between Boeing and Lockheed Martin established in 1996 to conduct the day-to-day operations of the Space Shuttle. It was based in Houston and had more than 10,000 employees at NASA's Johnson, Kennedy, and Marshal Centers.

Boeing, Nasa Systems
The space shuttle was designed and built by Rockwell International in Palmdale, California. In 1996 Boeing bought Rockwell International's assets related to the space program and later moved the company to Huntington Beach, California. Boeing was subcontracted by the United Space Alliance to carry out maintenance and modifications to the space shuttle.

Boeing, Rocketdyne Propulsion
Rockwell International's Rocketdyne Division was responsible for the space shuttle's main engines. It continued to support the engines when it became part of Boeing.

ATK Thiokol Propulsion
ATK Thiokol Propulsion manufactured the reusable solid propellant rockets in Bringham City, Utah.

Lockheed Martin, External Tanks
The external tanks were manufactured by Martin Marietta. They were later manufactured by Lockheed when the two companies merged to form Lockheed Martin. The external tank is the only disposable part of the space shuttle, so new tanks had to be produced constantly.

Lockheed Martin, Space Systems
The Reinforced Carbon Composite (RCC) panels used in the nose and leading edge of the space shuttle's wings were manufactured by Ling-Temco-Vought (LTV) in Grand Prairie, Texas. However, Lockheed Martin bought LTV after several acquisitions and mergers.

Facts about the Space Shuttle Program

The budget and the number of technicians and engineers involved in the Space Program gradually decreased throughout the 1960s and 1970s. Also, the NASA budget as a percentage of the Federal Budget had significantly cuts and from about 4% in the sixties it became roughly 1% in the seventies, eighties and nineties.

Put these facts together and we have an agency trying to do too much, with too few resources.

In June 1986, a Presidential Commission headed by William Rogers and with participants such as Neil Armstrong and Chuck Yeager handed to President Ronald Reagan a report on the Space Shuttle Challenger accident, which recommended measures be taken to increase the safety of the Space Program.

THE SPACE SHUTTLE

- Payload: 17.2 to 25.5 tons (depending on the orbit)
- 3 hydrogen/oxygen-powered main engines
- 2 solid propellant rocket engines (Solid Rocket Busters, SRB)
- 01 External Fuel Tank (Liquid Oxygen/Hydrogen, ET)

Space Shuttle[26]

Main Engines
The RS-25, the Main Engine (ME), was a cryogenic liquid fuel rocket engine used in NASA's Space Shuttle.

Built in the United States by Rocketdyne, the RS-25 burned liquid hydrogen and liquid oxygen cryogenic fuels, with each engine producing 1,859 kN (418,000 lbf) of thrust on take-off.

The RS-25 operated at extreme temperatures, with liquid hydrogen being stored at -250°C while the temperature in the combustion chamber peaked at 3,315°C.

External tank
The External Tank (ET) provided fuel for the main engines and structural support for attaching the solid-fuel rockets (SRBs) to the orbiter.

[26]https://spaceflight.nasa.gov/gallery/images/shuttle/sts-120/html/sts120-s-028.html. Public domain.

In the aft attachment area, umbilicals carried fluids, gases, electrical signals, and electrical power between the tank and the orbiter. Electrical signals and controls between the orbiter and the two solid fuel rockets were also routed through these umbilicals.

Solid Rocket Boosters (SRB)
They provided around 83% of the space shuttle's take-off thrust. Five of them fueled the first stage of the Saturn V rocket. The spent SRBs were recovered from the ocean, refurbished, recharged with propellant, and reused on various missions.

Flight controls

The space shuttle's flight control surfaces were:
- Rudder + Speed Brake
- Elevons
- Fuselage flaps
- Orbital maneuvering system
- Reaction flight control system

The engines of the reaction flight control system at the front of the fuselage were active until re-entry, and the engines of the reaction flight control system at the rear of the fuselage were deactivated when a specific dynamic pressure was detected.

The pitch motors of the rear reaction flight control system were deactivated at a dynamic pressure of 20 psi, when the vehicle's elevons became affective in pitch.

The roll motors of the rear reaction flight control system were deactivated at a dynamic pressure of 10 psi, when the vehicle's elevons became affective in roll.

The orbiter's speed brake was used below Mach 10. At approximately Mach 3.5, the rudder could be used, and the yaw motors of the rear reaction control system were deactivated.

THE CREW

Rick Husband, Commander. Husband, 45, was a USAF colonel, test pilot and STS-96 veteran. He received a bachelor’s degree in mechanical engineering from Texas Tech University and a master’s degree in mechanical engineering from California State University, Fresno.

William C. McCool, pilot. McCool, 41, was a US Navy commander and test pilot. He received a bachelor’s degree in applied sciences from the U.S. Naval Academy, a master’s degree in computer sciences from the University of Maryland, and a master's degree in Aeronautical Engineering from the U.S. Naval Postgraduate School.

Michael P. Anderson, payload mission and communications specialist. Anderson, 43, was a lieutenant colonel in the USAF, a former instructor pilot and tactical officer, and a veteran of STS-89. He received a bachelor's degree in physics and astronomy from the University of Washington and a master's degree in physics from Creighton University.

David M. Brown, mission specialist. Brown, 46, was a U.S. Navy captain, naval aviator, and naval flight surgeon. He received a bachelor's degree in biology from the College of William and Mary and a master's degree from Eastern Virginia Medical School.

Kalpana Chawla, Flight Engineer, and Mission Specialist. Chawla, 41, was an aerospace engineer, FAA-certified flight instructor, and STS-87 veteran. She received a bachelor's degree in Aeronautical Engineering from Punjab Engineering College, India, a master's degree in Aeronautical Engineering from the University of Texas, Arlington, and a doctorate in Aeronautical Engineering from the University of Colorado, Boulder.

Laurel Clark, mission specialist. Clark, 41, was a commander in the US Navy and a naval flight surgeon. She received a bachelor's degree in zoology and a doctorate from the University of Wisconsin, Madison.

Ilan Ramon, payload specialist. Ramon, 48, was a colonel in the Israeli Air Force, a fighter pilot, and Israel's first astronaut. Ramon graduated in Electronic and Computer Engineering from Tel Aviv University, Israel.

Crew Training

Mission training comprised 4,800 hours of training plus 3,500 hours for specific payload training. 16 integrated ascent and re-entry simulations were carried out.

THE MISSION

The SS Columbia was used for science missions because it was equipped for long-duration missions and was not being used for Shuttle-Mir docking missions or ISS assembly.

The NASA Administration recommended that NASA dedicate several Shuttle missions to microgravity and life sciences. The aim was to train scientists to take full advantage of the ISS's research capabilities once it became operational.

Mission highlights

- Experiments with the Advanced Respiratory Monitoring System stationary bike at SPACEHAB.

- Ozone Limb Sounding Experiment 2 began measuring the ozone layer.

- The Mediterranean-Israeli Dust Exp (MEIDEX) was created to measure atmospheric aerosols over the Mediterranean Sea and the Sahara Desert. MEIDEX captured images of dust clouds off the coasts of Nigeria, Mauritania, and Mali.

- European research into osteoporosis in space and on Earth. Experiments on the human body's response to microgravity continued, focusing on protein manufacture, bone and calcium production, kidney stone formation, and changes in saliva and urine due to viruses.

- Experiments with bone cells, prostate cancer, bacterial growth, thermal heating, and surface tension were planned.

A HAZARD THAT BECAME A REAL THREAT

Analysis showed that a large piece and at least two smaller pieces of insulating foam separated from the External Tank left bipod 81.7 seconds after launch.

Further analysis showed that the largest piece hit the Columbia in the lower part of the left wing.

The first formal meeting of the Debris Assessment Team was held five days after the start of the mission. It ended with the highest-ranking NASA engineer on the team agreeing to the team's request for images of the wing.

Members of the Debris Assessment Team later learned that managers had refused to take images of Columbia in orbit.

CONCLUSIONS OF THE ACCIDENT REPORT

1) In total, around seven pieces of foam larger than 2.5 centimeters in diameter hit the underside of the space shuttle during each mission since STS-6.

2) The physical cause of the loss of Columbia and its crew was the damage to the Thermal Protection System at the tip of the left wing. Tests with the systems installed on the space shuttle were able to reproduce equivalent damage on a test body.

RECOMMENDATIONS

1) To begin a program to remove all debris from the thermal protection of the external tank at source, with particular emphasis on the area where the bipod supports attach to the external tank.

2) To increase the shuttle's ability to withstand minor debris damage through more impact-resistant reinforced carbon-carbon (RCC) and protective plates. This program must determine the actual impact resistance of current materials and the effect of probable debris shocks.

3) To draw up an inspection plan to determine the structural integrity of all the components of the RCC system. This inspection plan must use non-destructive inspection technology.

4) For missions to the ISS, develop a practical capability to inspect and carry out emergency repairs on the widest possible range of damage to the Thermal Protection System, including plates and RCC, taking advantage of the availability of additional capacity when close to the ISS.

5) As far as possible, increase the shuttle's ability to successfully re-enter the Earth's atmosphere with little damage to the leading edge.

Columbia on the launch pad[27]

[27] The circle at the top of the image shows the bipod (in the shape of an ipsilon) at the rear attachment point. The circle at the bottom of the image shows the damaged RCC panel. *Source NASA Columbia Accident Report Volume 1.*

MAIN LESSONS LEARNED

Although many changes were made after the loss of Challenger, the organizational factors in the loss of Columbia were in many ways similar. Organization cultures are hard to change.

After the Cold War, NASA struggled to adapt to new political and financial challenges and persisted in pursuing its goals with diminishing resources.

"..the institutional practices that were in place at the time of the Challenger accident—such as failure to correct deviations from expected performance, a silent safety program, and schedule pressures—returned to NASA."

"..the process for resolving in-flight anomalies is not well documented and appears to change. Space Shuttle program managers seem to have confused the notion that foam poses a risk with foam not being a "flight safety issue." (Columbia Accident Investigation Board, CAIB)

"..when asked by investigators why they were not more vocal about their concerns, members of the Debris Assessment Team opined that by raising contrary views about the safety of the space shuttle mission, they would possibly be ridiculed by their colleagues and managers." (CAIB)

"The debris assessment team were in the unenviable position of wanting images to assess the damage more accurately and, at the same time, needing to prove to Program managers, as a result of their assessment, that there was a need for images in the first place." (CAIB)

AERONAUTICAL PRODUCT DEVELOPMENT

14

"Restrictions give directions to innovations."
Roni Gurwicz

PHASES OF DEVELOPMENT

The phases of Aeronautical Product Development can be defined as:

1) Conceptual Definition

Several possible configurations are studied according to mission, costumer, industrial, financial, logistics and certification requirements.

2) Initial Definition

The project has a "go ahead" to proceed. An initial configuration is defined.

3) Detailed Definition

The potential suppliers are contacted, and further studies are made to define the complete design, supply chain, production and all logistic aspects of the project.

4) Joint Definition

The design reaches an acceptable configuration and all suppliers are defined. All the mission, costumer and certification requirements are verified.

5) Testing

Numerous tests are made to verify airworthiness of the design and to verify compliance with requirements.

6) Certification

Certification tests and test results involving an Aviation Authority to verify compliance with certification requirements.

7) Entry-into-Service

Costumers start to receive the aircraft, training, parts etc.

8) Product Support

The costumers are supported by the manufacturer during service with technical assistance and sometimes logistic support.

9) Phase out

The aircrafts are continuously removed from service.

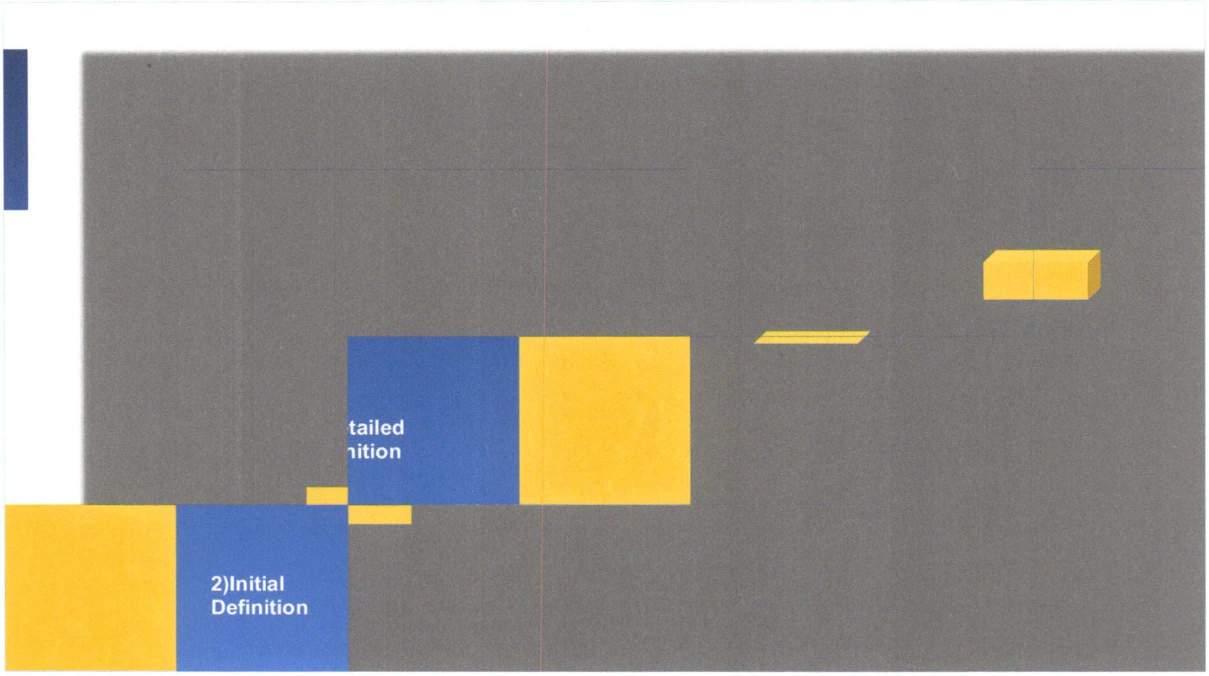

The Phases of Aircraft Development

ACTIVITIES DURING PRODUCT DEVELOPMENT

Throughout the phases of Product Development, certain activities are carried out repeatedly and many are related. Didactically, the activities that take place during these development phases can be grouped into one of the five following groups:
- Resources management
- Integration activities
- Testing and validation activities
- Results management
- Threats management

Resource management involves material and human resources (proprietary and from suppliers). The identification of material needs time and continuous planning.

Integration activities are all activities that modify the product or production process. This can also include resources necessary for future integration, such as the construction of rigs, design changes, etc.

Testing and validation activities are those activities that verify if the integration activities have been satisfactory.

Results management is the activity that monitors and formalizes the results obtained.

Risk management involves monitoring possible undesirable outcomes for the company and taking actions to make sure they do not occur.

DESIGN REVIEWS

These phases are usually carried out in sequence, one after the other, but there are periods when they occur simultaneously. To show the end of one phase and the start of another and assess the progress of the project, Phase Verification meetings are held, at which deliverables are defined and presented. Deliveries are usually divided by area and each area presents the status of its deliveries and planning.

The Phase Verification meetings are generally called "Design Reviews". The Design Reviews are also called according to the phase they are related to:
- Preliminary Design Reviews (PDR)
- Detailed Design Reviews (DDR)
- Critical Design Reviews (CDR)

In addition, possible risks at the end of the phase and actions to mitigate these risks are investigated. These risks are possible sources of costs, delays, and redesigns. Phase Checks, therefore, seek to identify them.

Difficulties encountered during design reviews

Phase Checks are generally effective, but may not work properly due to various problems:
- Failures or the non-existence of previous Phases (more frequent than one can imagine)
- Deliverables not clearly defined
- Deadlines not clearly defined
- Technical issues with the product
- Corporate culture
- Insufficient resources (material or human)
- Lack of preparation

It is also true that the later any problems are dealt with, the more difficult and expensive it becomes to solve them. If a particular correction is identified late, for example in the Entry-into-Service phase, it will lead to exponentially greater delays and costs than if it is identified and corrected at an earlier stage. This is why it is essential to identify problems as early as possible. Therefore, it is essential to test the aircraft as early as possible.

- Long duration: Phase Checks usually last between two to five days, making it practically impossible to participate in all the forums. The results are, therefore, restricted to a small number of people. It can happen that the people who have access to the information are not the ones who need it.

- Not using previous results: previous results must be added to current results to ensure that all issues are covered.
- Use of inappropriate tools: tools generally accepted and used in Phase Checks, such as MS PowerPoint presentations, are easy to prepare, but often hide the issues over an excessive number of slides.
- Failure to incorporate the conclusions and lessons learned into regular activities: the exercise of monitoring threats must be carried out continuously to prevent them from spreading.

RISKS

Risks are a major cause of delays and costs.

FLIGHT TEST RISKS

The Flight Test risk matrices must map all the foreseeable "top events" of the Flight Test Campaign for each Flight Test Plan. Each Test Plan must have its risks assessed according to the risk methodology used by the company.

PROGRAM RISKS

The Program's risks must map out all the foreseeable "top events" for each technology involved. The risks must be assessed according to the risk methodology used by the company. Normally, the risks of the program are assessed according to the probability of occurrence and possible consequences.

AERONAUTICAL DESIGN ORGANIZATIONS

15

"If you run to judge, you don't have time to understand."
Yuval Harari

WHY DESIGN ORGANIZATIONS

Design Organizations are those that have the prerogatives, accepted by the Certifying Authorities, to carry out aeronautical designs and modifications to aeronautical designs. The applicable requirements for obtaining certification as a Design Organization are described in FAR or CS 21 subpart J.

These organizations have the right to request Type Certificates (TC) for their projects from the authorities. They also have the right to request modifications to these Type Certificates. And they also have the prerogative to carry out major repairs.

For an organization to be accepted as an Aeronautical Design Organization, it has to comply with the requirements of FAR/CS 21 Subpart J. To do so, it needs to have its own means of developing products and verifying and demonstrating that these products meet the design requirements. These means must be accepted by the aeronautical authority of the country it operates in.

The Design Organization must ensure that:

1. The technical departments have enough people and have the appropriate experience and authority to be able to carry out the responsibilities assigned to them.

2- Facilities, equipment, and accommodation are adequate to enable personnel to comply with airworthiness and operational suitability data and environmental protection requirements relating to the product.

3. There is full and efficient coordination between departments and within departments concerning airworthiness, operational suitability data, and environmental protection issues.

Flight Test Organizations may or may not be part of a Design Organization.

THE DESIGN MANAGEMENT SYSTEM

The Design Organizations need to establish, implement, and maintain a process called the "Design Management System." This system aims to ensure that it can:
- Design products that comply with certification requirements.
- Verify compliance with these requirements.
- Demonstrate compliance with these requirements to the certification authority.

The Design Management System must:

- Match the size of the organization and the nature and complexity of its activities, considering the hazards and risks associated with these activities.

- Be created and maintained under the responsibility of a single manager appointed according to paragraph 21.A.245(a) of the Civil Aviation Regulations.

O sistema de Gerenciamento de Projeto tem duas partes:
- Garantia de Projeto
- Gerenciamento da Segurança

THE HEAD OF THE DESIGN ORGANIZATION

A head of the design organization must be appointed. In turn, the head of the Design Organization must appoint and specify the extent of their authority to:

1. A head of the Airworthiness function
2. A head of the Independent Monitoring function
3. Depending on the size of the organization and the nature and complexity of its activities, any other person or group of persons necessary to ensure that the organization complies with the requirements of Subpart J of the Civil Aviation Regulations

DESIGN ORGANIZATION HANDBOOK

The Design Management System must also have a Handbook per CS 21.A.243, which must be delivered to the Aviation authority.

The following is a typical table of contents for a Design Organization Handbook:

Part 1: Organization

- 1.1 Objective of the manual
- 1.2 Responsible for administering the manual
- 1.3 Amendment procedures
- 1.4 List of effective pages
- 1.5 Distribution list
- 1.6 Presentation of the Design Organization (DO)
- 1.7 Scope of work (with identification of product types and models)
- 1.8 Organizational charts
- 1.9 Human resources
- 1.10 Management Staff
- 1.11 Certification staff
- 1.12 System-independent monitoring
- 1.13 Safety management system

Part 2: Procedures

- 2.1 Management of changes to the type design and repair design:

- Configuration control,
- Classification, and
- Approval of minor changes to the type design and minor repairs.

2.2 Control of project subcontractors
2.3 Collecting/Investigating faults, malfunctions and defects
2.4 Coordination with production
2.5 Documentation control:
- Regarding alterations and repairs, and
- Regarding faults, malfunctions and defects (i.e. service bulletins)

2.6 Record keeping

PARTS OF A DESIGN ORGANIZATION

A Design Organization has four parts which are described in the following paragraphs.

Parts of a Design Organization

DESIGN MANAGEMENT SYSTEM

It is the core of a Design Organization. To have a Design Management system, the Design Organization must:

- Establish, implement, and maintain a system for the control and supervision of the design, design changes and repairs of products, parts, and appliances covered by the terms of approval.

- Specify how to assess the acceptability of the parts or equipment designed or the tasks carried out by its partners or subcontractors, following written methods and procedures.

If the design organization holds one or more additional organization certificates under Regulation (EU) 2018/1139, the design management system may be integrated with that required by the additional certificate(s).

INDEPENDENT MONITORING

The independent monitoring function aims to verify the organization's compliance with the relevant requirements, as well as the compliance and adequacy of the Design Management system. Monitoring shall include *feedback* to the person or group of persons referred to in 21.A.245(b) and to the manager referred to in 21.A.245(a) to ensure, where necessary, the implementation of corrective measures.

AIRWORTHINESS

A Design Organization shall have an Airworthiness function responsible for ensuring that the design of products, parts, and appliances, or design changes and repairs, comply with the applicable type-certification basis, the applicable operational suitability data certification basis, and environmental protection requirements.

SAFETY MANAGEMENT

As part of the safety management element the design organization must:

1. Establish, implement, and maintain a safety policy and corresponding safety objectives.

2. Appoint key safety personnel per 21.A.245(b).

3. Establish, implement, and maintain a safety risk management process that includes the identification of risks to aviation safety arising from its activities, their assessment, and the management of associated risks, including the adoption of actions to mitigate the risks and verify their effectiveness.

4. Establish, implement, and maintain a safety process that includes:

(i) measuring and monitoring the organization's safety performance
(ii) the management of changes per points 21.A.243(c) and 21.A.247
(iii) the principles of continuous improvement in Safety Management.

5. Promote safety in the organization by:

(i) training
(ii) education and
(iii) communication

6. Establish an occurrence reporting system per 21.A.3A to contribute to the continuous improvement of safety.

Safety should not be considered the responsibility of a single person or a limited group of people in the organization. A safety culture must be developed throughout the organization, involving all personnel as active contributors to the safety of the final product or its parts.

Safety Manager

If more than one person is assigned to the development, administration, and maintenance of effective safety management processes as defined in 21.A.239(c)(2), the Head of the Design Organization shall identify the safety manager as the single focal point.

The role of the safety manager should be to:
(1) Facilitate the identification of hazards and the assessment and management of risks
(2) Monitor the implementation of the actions taken to mitigate the risks, as listed in the safety action plan, unless the follow-up of the actions is done by the independent monitoring function.
(3) Provide periodic reports on safety performance to the safety review board (the functions of the safety review board are defined in AMC1 21.A.239(c)(2)).
(4) Ensure the maintenance of safety management documentation.
(5) Ensure that safety training is available and meets acceptable standards.
(6) Provide advice on safety issues.
(7) Ensure the initiation and follow-up of internal investigations into incidents.

AERONAUTICAL PRODUCTION ORGANIZATIONS

16

"Strategy is to see distant things as if they were close and close things as if they were distant."
Myamoto Musashi

WHY PRODUCTION ORGANIZATIONS

Aeronautical Production Organizations are a type of organization that has been recognized by the Aeronautical Authority of a country as competent to produce aircraft or parts according to design data. Recognition of an organization as a Production Organization follows the provisions of FAR/CS 21 Part G.

Subpart G of CS/FAR 21 provides the rules for demonstrating compliance of products, parts, and devices with applicable design data through an approved Design Organization.

A Production Organization Exposition (POE) must be made and there must be a link between a Design Organization and the Production Organization.

PRODUCTION ORGANIZATION EXPOSITION (POE)

The Production Organization shall establish and maintain a Production Organization Exposition Manual ("POE") that provides the following information related to the Organization, as described in paragraph 21.A.139 of the requirements:

1. A statement signed by the responsible manager confirming that the Production Organization Exposition, and any associated manuals that define the organization's compliance with the requirement, will always be complied with.

2. Title(s) and names of officers accepted by the authority per paragraph 21.A.145(c)(2) of this standard.

3. The duties and responsibilities of the manager(s), as required by paragraph 21.A.145(c)(2), including matters on which they may deal directly with the Aviation Authority on behalf of the organization.

4. An organizational chart showing the associated chains of responsibility of managers, as required by paragraph 21.A.145(c)(1) and (2) of the standard.

5. A list of the certification personnel referred to in paragraph 21.A.145(d).

6. A general description of human resources.

7. A general description of the facilities located at each address specified on the production organization's certificate of approval.

8. A general description of the scope of work of the production organization relevant to the terms of approval.

9. The procedure for notifying organizational changes to the Aviation Authority.

10. The procedure for amending the Production Organization Exposition Manual ("POE").

11. A description of the Production Management System and the policy, processes, and procedures set out in paragraph 21.A.139(c).

12. A list of the external parties referred to in paragraph 21.A.139(d)(1).

The initial issue of the "POE" must be approved by the Aviation Authority.

The "POE" should be amended as necessary so that it is an up-to-date description of the organization. Copies of any changes must be provided to the Aviation Authority.

PARTS OF A PRODUCTION ORGANIZATION

Similarly to the Design Organizations, the Production Organizations have four parts according to the image below.

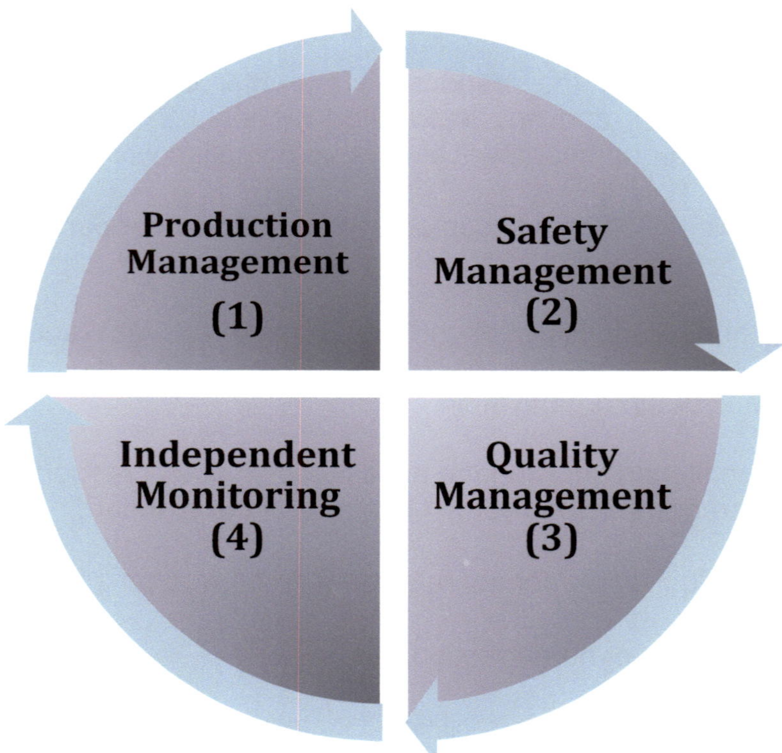

Parts of a Production Organization

PRODUCTION MANAGEMENT SYSTEM

The production organization must establish, implement, and maintain a production management system to clearly defined responsibilities and lines of accountability throughout the organization.

The production management system must:

1. Match the size of the organization and the nature and complexity of its activities, taking into account the hazards and risks associated with these activities.

2. Be created, implemented, and maintained under the direct responsibility of a single manager appointed per 21.A.145(c)(1).

SAFETY MANAGEMENT SYSTEM

As part of the safety management element, the Production Organization must:

1. Establish, implement, and maintain a safety policy and the corresponding safety objectives.

2. Appoint key safety personnel per paragraph 21.A.145(c)(2).

3. Establish, implement, and maintain a safety risk management process to identify the safety risks arising from its aviation activities, assess them, and manage the associated risks, including the adoption of actions to mitigate the risks and verify their effectiveness.

4. Establish, implement, and maintain a Safety Assurance process that includes:

    (i) measuring and monitoring the organization's safety performance.

    (ii) the management of changes under paragraph 21.A.147.

    (iii) the principles for continuous improvement of the safety management system.

5. Promote safety in the organization by:

    (i) training
    (ii) education and
    (ii) communication

6. Establish an occurrence reporting system per paragraph 21.A.3A to contribute to the continuous improvement of safety.

QUALITY MANAGEMENT SYSTEM

As part of the Quality Management element, the Production Organization must:

1. Ensure that each product, part, or appliance produced by the organization or its partners, suppliers or subcontractors, complies with the applicable design data and is in a condition for safe operation, thus allowing the exercise of the privileges indicated in paragraph 21.A.163.

2. Establish, implement, and maintain, as appropriate, within the scope of the approval, control procedures for:

(a) issuing, approving, or amending documents.

(b) audit and evaluation of suppliers and subcontractors.

(c) verification that the products, parts, materials, and equipment received, including items supplied new or used, are as specified in the applicable design data.

(d) identification and traceability.

(e) manufacturing processes.

(f) inspection and testing, including production flight tests.

(g) calibration of tools, jigs, and test equipment.

(h) control of non-conforming items.

(i) airworthiness coordination with the applicant or holder of the design approval.

(j) filling in and retaining records.

(k) staff competence and qualifications.

(l) issuing airworthiness documents.

(m) handling, storage, and packaging.

(n) internal quality audits and resulting corrective actions.

(o) work carried out at any location other than the approved premises.

(p) work carried out after completion of production, but before delivery, to maintain the aircraft in a safe operating condition.

(q) issuance of flight authorization and associated flight approval or conditions.

3. Include specific provisions in the control procedures for any critical parts.

INDEPENDENT MONITORING SYSTEM

The production organization shall establish an independent monitoring function to verify the organization's compliance with the relevant requirements, as well as the compliance and adequacy of the Production Management System.

Monitoring shall include feedback to the person referred to in paragraph 21.A.145(c)(2) and to the manager referred to in paragraph 21.A.145(c)(1) to ensure, where necessary, that corrective action is implemented.

If the production organization holds one or more additional organization certificates under Regulation (EU) 2018/1139, the Production Management System can be integrated with the required by the additional certificate obtained.

LINK BETWEEN DESIGN AND PRODUCTION ORGANIZATIONS

When the design and production organizations are two separate legal entities, a Delivery Authorization must be available for direct delivery to end users to ensure control of the airworthiness of parts and equipment.

The organization responsible for production and the one responsible for the design must both work as a consolidated team.

For an organization to be eligible to become a Production Organization it must:

(a) justify that, for a defined scope of work, approval under Subpart G is appropriate for demonstrating compliance with a specific project.

(b) have, or have applied for, approval of that specific project; or

(c) have ensured satisfactory coordination between production and the design (done via an agreement). An agreement is considered satisfactory if it is documented and satisfies the authority that coordination is satisfactory.

To achieve satisfactory coordination, the agreements must define at least the following aspects, regardless of whether the two organizations are separate legal entities or not:

- The responsibilities of a Design Organization that ensure the correct and timely transfer of airworthiness data (e.g. drawings, material specifications, dimensional data, processes, surface treatments, transport conditions, quality requirements, etc.).

- The responsibilities and procedures of the POA holder or applicant to develop, where applicable, its own manufacturing data per the airworthiness data package.

- The responsibilities of a POA holder or applicant to assist the design organization in addressing continuing airworthiness issues and taking the necessary actions (e.g. traceability of parts in case of direct delivery to users, installation of modifications, traceability of process results and approved deviations for individual parts as applicable, technical information, etc.).

- The agreements should cover the requirements of part 21, subpart G, and associated AMC and GM, in particular: 21.A.145(b), 21.A.165(c), (f) and (g).

- The responsibilities of a POA holder or applicant, before type certification, to help a design organization demonstrate compliance with the requirements (access to and suitability of production and testing facilities for the manufacture and testing of prototypes and test specimens).

- The procedures for dealing with production deviations and non-conforming parts.

- The procedures and responsibilities for configuration control of manufactured parts, to enable the Production Organization to make the final determination and identification for compliance or release of airworthiness and eligibility status.

- The identification of the responsible persons or offices who control the above.

- Acknowledgement by the holder of the TC/STC/repair approval or TSO change/authorization that the approved design data provided, controlled, and modified by the agreement is recognized as approved.

In many cases, the Production Organization may receive the approved design data through an intermediate Production Organization. This is acceptable if an effective link can be maintained between the design approval holder and the production organization to satisfy the intent of 21.A.133.

FINAL WORDS

17

"This is not a drill."
Roger Waters

It is common for a professional to have gaps in their experience that take a long time to fill. In the case of flight test professionals, sometimes this gap is in technical knowledge, sometimes in operational knowledge, and sometimes in the way a Flight Test Organization should operate so that it is in harmony with the rest of the company and produces maximum results. Chapters 1 to 9 of this book aim to fill some of the gaps related to organizational knowledge.

In addition, complex processes involving many people tend to leave "dormant gaps" that can generate unpredictable and undesirable redesigns, risks, and costs.

Many of these failures end up being dealt with in flight tests because they are the aircraft's first operators, when the systems are still not very mature and when there are still many uncertainties about the aircraft's capabilities. For this reason, good knowledge of risk management becomes important and is covered in Chapters 10 to 13.

Therefore, a broad view of the processes and how an aeronautical product should be developed is very important for flight test professionals. Chapters 14 to 16 aim to provide this insight.

The centuries-old Japanese art of archery, called Kyudo, has many practical teachings. The most important thing for them is not to hit the center of the target, but the process used to launch the arrow. The result is not the most important thing, but the way you handle the task.

The knowledge presented in this book cannot simply be learned rationally, it must be experienced and understood. This is a long process and how it is handled is more important than the result, just as in the art of archery.

To those of you who have made it this far, I would like to express my sincere thanks, and I hope that this book will be a useful tool and will help to provide a very happy professional experience for everyone who reads it.

Maximilian Kleinubing
September 06, 2024

Progress always occurs outside of the comfort zone.
Michael John Bobak

REFERENCES

18

"An old owl lived in an oak.
The more he saw the less he spoke.
The less he spoke the more he heard.
Why aren't we all like that wise old bird?"
John D. Rockfeller

1. Columbia Accident Board Investigation Report, Columbia Accident Investigation Board, National Aeronautics and Space Administration and the Government Printing Office, Report Volume 1, August 2003.
2. http://en.wikipedia.org/wiki/Space_Shuttle_Main_Engine
3. http://en.wikipedia.org/wiki/Space_Shuttle_Solid_Rocket_Booster
4. https://www.officialdata.org/us/inflation/
5. HURT, H. H Jr. Aerodynamics for Naval Aviators, US Naval Air Systems Command, NAVAIR Report #00-80T-80, Revised January 1965.
6. G650 Flight Test Accident Gulfstream Party Submission, DCA11MA076, May 21, 2012.
7. Gulfstream G650 NTSB Preliminary Report, site: www.ntsb.gov.
8. Introduction to Flight Test Engineering, NATO Research and Technology Organization, RTO-AG-300-V14, Flight Test Technique Series Vol.14, North Atlantic Treaty Organization (NATO), France, 2005.
9. Basic Principles of Flight Test Instrumentation Engineering, Advisory Group for Aerospace Research and Development, AGARD-AG-160 Vol.1 (Issue 2), North Atlantic Treaty Organization (NATO), Neuilly Sur Seine, France, 2005.
10. Grob Aerospace G180A Accident Report, Bundesstelle für Flugunfalluntersuchung, 2006.
11. WHITTLE F. The early history of the Whittle jet propulsion gas turbine, Institution of Mechanical Engineers, James Clayton Memorial Lecture, London, 1945.
12. Air Transport Association of America, Maintenance Program Development Document, MSG-3, 1993.
13. Yeager An Autobiography, Yeager C. and Janos, L., Bantam Books, New York, 1985.
14. I Could Never be so Lucky Again, Doolittle J. and Glines C. V., Bantam Books, New York, 1991.
15. The Untouchables, Shul Brian and Watson Walter, Mach 1 Inc. 1994.
16. REASON, James. "The Contribution of Latent Human Failures to the Breakdown of Complex Systems," 1990.
17. Lilienthal, Otto, Der Vogelflug als Grundlage der Fliegekunst Ein Beitrag zur Systematik der Flugtechnik, Berlim, 1889.
18. U.S. Department of Transportation, Federal Aviation Administration, Aircraft Certification Service Flight Test Risk Management, FAA Order 4040.26C, 2021.
19. European Union Aviation Safety Agency, Acceptable Means of Compliance (AMC) and Guidance Material (GM) to Annex I (Part 21) to Commission Regulation (EU) No 748/2012, Issue 2, Amendment 16, 2023.
20. ICAO International Standards and Recommended Practices, Annex 19 to the Convention on International Civil Aviation, Safety Management, second edition, July 2016.

www.ingramcontent.com/pod-product-compliance
Lightning Source LLC
Chambersburg PA
CBHW041324210926
51847CB00003B/37
* 9 7 8 6 5 0 1 6 2 3 8 0 1 *